Linear Audio - *Volume 0*

Jan Didden

publisher / editor

Authors:

Frank Blöhbaum

Andy Bryner

Bob Cordell

Jean-Claude Gaertner

Joachim Gerhard

Siegfried Linkwitz

Tom Nousaine

Nelson Pass

Ari Polisois

Douglas Self

Ed Simon

Pierre Touzelet

René Wouda

Stuart Yaniger

Linear Audio Volume 0 is published by Linear Audio of Hoensbroek, The Netherlands.
Published in September 2010.

Important notices and disclaimers

Many of the designs, projects, products, methods, instructions or ideas in Linear Audio publica-tions involve potentially lethal voltages. The information is provided with the understanding that the practitioner has sufficient knowledge and experience to safely execute the ideas and practices described in this publication, and to avoid injury and damages to persons and property. Safe and professional practices are assumed. If you are not sure to have this knowledge and experience, you should seek experienced professional advice and assistance.

Neither the Publisher nor the authors, contributors or editors assume any liability for any injury and/ or damage to persons or property as a matter of product liability, negligence or otherwise, or from any use or operation of any projects, products, methods, or ideas contained in the material in Linear Audio publications and materials.

The designs, projects, products, methods, instructions or ideas in Linear Audio publications may be subject to patents, trademarks or similar deposits, which are all recognized.

The use of general descriptive or registered names, trademarks or similar identifications in this pub-lication does not imply that these are free for general use.

Design, Prepress: Isolde Philips - www.isoldephilips.nl
Production: F&N Eigen Beheer, Amsterdam, The Netherlands
ISBN: 9 789490 929015

www.linearaudio.net

From the editor's desk

Most historians agree that the modern printing press was invented in 1440 by Johannes Gutenberg. Now, almost 600 years later, some people feel that the printing process is gradually but inexorably being replaced by digital media distribution.

I do not agree.

Just like MP3 did not kill good, up-front stereo music reproduction, so does digital distribution not kill distribution of printed works. This bookzine is proof of that.

And it is not just a matter of what you grow up with. Even the highest resolution digital material on wide screens cannot replace the joy of holding a book, smelling a book and leafing through a book, picking up text and graphic fragments here and there.

And in case of an audio publication, nothing can replace the book on your workbench, opened at your current project description while you are working to bring that circuit to life, or while you are working to make sense of those speaker measurements.

Likewise, when later, in a relaxed mood, you get a flash of insight and grab that book and check those numbers and eureka!

MP3 is complementary to traditional 2-way stereo or surround sound. Digital media distribution is complementary to real-life, hands-on printed matter.

Audio has been my vocation as long as I can remember, and indeed I cannot imagine a life without being involved in audio engineering and music. Retirement for me was just a great opportunity to devote more time to audio. Regrettably, audio periodicals dwindled in numbers over the last few decades. Those that survive do so by the total dedication of a few stubborn individuals. I wish to publicly acknowledge here my very great admiration and respect for Edward T Dell Jr, who founded The Audio Amateur in 1970, which today is the only dedicated audio periodical in existence, published as AudioXpress. All of us who are into audio either as a hobby or professionally owe him more than we can imagine. Personally I owe Ed even more as he has encouraged me to write articles and interviews; he knew better what I would enjoy than I did! Without him, I would have written much less than I have, and this bookzine would not exist.

These days I spend many an hour online, especially at diyaudio.com. This place has to be experienced to be believed. It is huge, but most people have their own niche where they meet audio soul

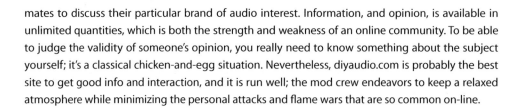

mates to discuss their particular brand of audio interest. Information, and opinion, is available in unlimited quantities, which is both the strength and weakness of an online community. To be able to judge the validity of someone's opinion, you really need to know something about the subject yourself; it's a classical chicken-and-egg situation. Nevertheless, diyaudio.com is probably the best site to get good info and interaction, and it is run well; the mod crew endeavors to keep a relaxed atmosphere while minimizing the personal attacks and flame wars that are so common on-line.

By now you probably have a hunch that the above seemingly unrelated tangents are not unrelated at all. Fact is that each of the three situations mentioned above eventually drove me to the adventure of publishing Linear Audio.

I want to contribute to the availability of printed material with interesting audio-related articles. I want to reinforce the availability-base of periodic, good quality audio construction and educational articles. I want to provide a complement to on-line audio communities with a medium where a subject can be treated in-depth and less fleeting, based on facts and figures.

There you have it; both the bookzine and my reasons for getting it out to you. It is a very enjoyable experience for me to work with the authors, to share their enthusiasm for bringing you these interesting articles. Please visit www.linearaudio.net to read about the authors, their background and their involvement in audio.
I hope; no I'm certain, that you will enjoy this bookzine. There's such a wealth and wide range of articles, touching on many areas of audio! My wish is that it will motivate you to finally take up that project you always wanted to do and experience the joy of a job well done. Homo Sapiens has become Homo Faber, man the maker. Enjoy, and please let me know what you like, what you dislike, and what you want to see in a future issue. There's a place on the website where you can leave your feedback.
And if you have an article or article idea, by all means let me know!

Jan Didden
Linear Audio
Publisher/Editor

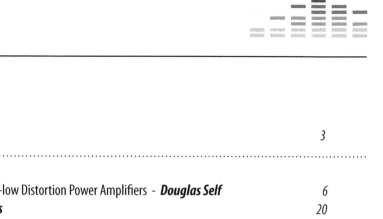

From the Editor's desk   *3*

Power Amplifiers
Inclusive Compensation & Ultra-low Distortion Power Amplifiers - **Douglas Self**   *6*
The Arch Nemesis - **Nelson Pass**   *20*

PreAmplifiers
Down the Rabbit Hole - Adventures in the Land of Phonostages - **Joachim Gerhard**   *31*

Tubes
A New Low-Noise Circuit Approach for Pentodes - **Frank Blöhbaum**   *46*
The Mini-Simplex - **Ari Polisois**   *66*
Split the Difference: The Truth about the Humble Cathodyne - **Stuart Yaniger**   *78*

Loudspeakers
STEREO - From live to recorded and reproduced - What does it take? - **Siegfried Linkwitz**   *89*
Project 21 Part I - The Satellites - **Jean-Claude Gaertner**   *112*
Low frequencies in rooms - **Tom Nousaine**   *138*

Test Equipment
The Distortion Magnifier - **Bob Cordell**   *142*

Transformers
On the leakage inductance in audio transformers - **Pierre Touzelet**   *151*

Tips & Tricks
The IC holder notebook - **Ed Simon**   *162*

Book Review
Douglas Self, Small Signal Audio Design - **Andy Bryner**   *164*
Bob Cordell, Designing Audio Power Amplifiers - *sneak preview* - **ed.**   *167*

Musings
Remember your first single? - **René Wouda**   *168*

Inclusive compensation & Ultra-low distortion power amplifiers

Douglas Self

Two things define the distortion performance of a power amplifier: the open-loop linearity of the circuitry, and the amount of negative feedback that can be safely applied. The latter is determined by the compensation scheme used. The almost-universal Miller dominant-pole method gives excellent and reliable results, but can be significantly improved upon by techniques such as two-pole compensation [1] which allow more feedback at audio frequencies. Inclusive compensation promises even better performance, but as normally conceived it is not workably stable. Here I show how to make it work. The results are dramatic, and the extra cost is trivial.

The vast majority of audio power amplifiers have three stages: a differential input stage that performs voltage-to-current conversion (ie, a transconductance stage), a voltage-amplifier stage (VAS) that performs current-to-voltage conversion, and a unity-gain output stage with a high input impedance but large current-output capability.

A power amplifier must be compensated because its open-loop gain is usually still high at frequencies where the internal phase-shifts add up to 180°. This turns negative feedback into positive at high frequencies, and causes oscillation, which can be very destructive to both the amplifier and attached loudspeakers. This dire scenario is prevented by adding some form of HF roll-off so the loop gain falls to below unity before the phase-shift reaches 180°, so oscillation cannot develop. This process of Compensation makes the amplifier stable, but the way in which it is applied has major effects on its closed-loop distortion performance.

Conventional Miller compensation

Almost all audio power amplifiers use dominant-pole compensation. A "pole" here is a frequency from which the open-loop gain falls of at 6 dB/octave; when a second pole at a higher frequency comes into action, the gain falls off at 12 dB/octave. To implement dominant-pole compensation, you take the lowest frequency pole that exists and make it dominant; in other words so much lower in frequency than the next pole up that the total loop-gain (i.e., the open-loop gain as reduced by the attenuation in the feedback network) falls below unity before enough phase-shift builds up to

cause HF oscillation. With a single pole, the gain must fall at 6 dB/octave, corresponding to a constant 90° phase shift. Thus the phase margin, the difference between the actual phase shift and the 180° that will cause oscillation is 90°, which gives good stability.

Since we have caused the open-loop gain to roll-off, the amount of feedback available to linearise the amplifier is reduced at high frequencies, and distortion will be greater. This disadvantage can be offset by applying the dominant-pole in the form of Miller feedback [2] from collector to base of the VAS; this is Ccompen in Figure 1. This is by far the commonest method of dominant-pole compensation, but it is also the best. Its action is in fact rather subtle. As the frequency increases, the local feedback around the VAS through the Miller capacitor also increases, so that open-loop gain which was being applied for global negative feedback smoothly transforms itself into gain that linearises the VAS only. Since this stage has to generate the large voltage-swing required to drive the unity output stage, maintaining its linearity in this way is a thoroughly good idea.
I am using the term "local feedback" to refer to feedback that is confined to a single stage only, such as the Miller capacitor around the VAS; other examples are emitter degeneration resistors in an input differential stage and the use of Complementary Feedback Pairs (CFP) in an output stage.

There are two other advantages to Miller dominant-pole compensation. Firstly, the input impedance of the VAS is lowered by the local shunt feedback through the Miller capacitor, so the VAS input is effectively a virtual-earth current-input; this prevents unwanted roll-offs occurring at the output of the input stage. Secondly, this local feedback reduces the output impedance of the VAS, enabling it to effectively drive the non-linear input impedance of the output stage.

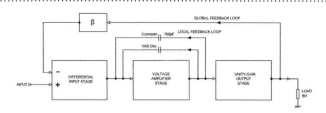

Figure 1   The feedback paths in a conventional Miller-compensated amplifier.

Figure 1 shows the feedback paths in a normal Miller dominant-pole compensated power amplifier. The block marked β represents the attenuation in the global feedback path required to give a closed-loop gain of 1/ β times. Also shown is the collector-base capacitance Cbc, which exists inside the VAS transistor and is non-linear, being a function of the collector voltage Vce. In a typical high-voltage small-signal transistor such as the MPSA42, often used in a VAS, it is quoted as a maximum of 3.0 pF for Vce = 20V, so the normal Miller value of around 100 pF is enough to swamp it. It is included here in every diagram as a reminder that whatever fancy compensation schemes you dream up, this capacitance must be taken into account; if necessary it could be rendered harmless by using a cascoded VAS. The non-linearity of Cbc is not an issue in the input stage transistors as there is only a very low signal voltage on their collectors.

This form of amplifier can give a very good distortion performance, as described below in the section on the Blameless amplifier philosophy, but the great limitation on this is the high-order distortion products generated by crossover distortion in the output stage. These are at a high frequency and so are less effectively linearised by the global feedback factor, which falls with frequency due to the dominant-pole compensation.

This brief description hopefully demonstrates why Miller dominant-pole compensation is near universal; it positively bristles with advantages, is cheap to implement, and gives thoroughly dependable performance. Any alternative compensation scheme has to improve on this. It is a tall order.

Output-inclusive compensation

Looking at Figure 1, we note that the output stage has unity gain, and it has occurred to many people that the open-loop gain and the feedback factor at various frequencies would be unchanged if the Miller capacitor Ccompen was driven from the amplifier output, as in Figure 2, creating a semi-local feedback loop enclosing the output stage. "Semi-local" means that it encompasses two stages, but not all three- that is the function of the global feedback loop. The inclusion of the output stage in the Miller loop has always been seen as a highly desirable goal because it promises that crossover distortion from the output stage can be much reduced by giving it all the benefit of the open-loop gain inside the Miller loop, which does not fall off with frequency until the much smaller Cbc takes effect.

Figure 2 shows the form of output-inclusive compensation that is usually advocated, and it has only one drawback- it does not work. I have tried it many times and the result was always intractable high-frequency instability. On a closer examination this is not surprising.

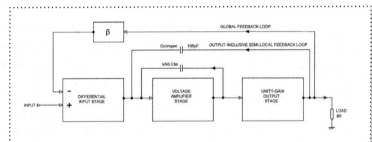

Figure 2   The feedback paths in an output-inclusive compensated amplifier. Note that the collector-base capacitance Cbc is still very much present.

Problems of inclusion

Using local feedback to linearise the VAS is reliably successful because it is working in a small local loop with no extra stages that can give extra phase-shift beyond that inherent in the Ccompen dominant pole. Experience shows that you can insert a cascode or a small-signal emitter-follower

into this loop, [3] but a slow output stage with all sorts of complexities in its frequency response is a very different matter. Published information on this is very scanty, but Bob Widlar [4] stated in 1988 that output stage behaviour must be well-controlled up to 100 MHz for the technique to be reliable; this would appear to be flat-out impossible for discrete power stages, made up of devices with varying betas, and driving a variety of loads.

Trying to evaluate what sort of output stage behaviour, in particular frequency response, is required to make this form of inclusive compensation workably stable quickly runs into major difficulties. The devices in a typical Class-B output stage work in voltage and current conditions that vary wildly over a cycle that covers the full output voltage swing into a load. Consequently the transconductances and frequency responses of those devices, and the response of the output stage overall, also vary by a large amounts.

The output stage also has to drive loads that vary widely in both impedance modulus and phase angle. To some extent the correct use of Zobel networks and output inductors reduces the phase angle problem, but load modulus still has a direct effect on the magnitude of currents flowing in the output stage devices, and has corresponding effects on their transconductance and frequency response.

Input-inclusive compensation

An alternative form of inclusive compensation that has been proposed is enclosing the input stage and the VAS in an inner feedback loop, leaving just the output stage forlornly outside. This approach, shown in Figure 3, has been advocated many times, one of its proponents being the late John Linsley-Hood, for example in [5]. A typical circuit is shown in Figure 4. It was his contention that this configuration reduced the likelihood of input-device overload (ie slew-limiting) on fast transients because current flow into and out of the compensation capacitor was no longer limited by the maximum output current of the input pair (essentially the value of the tail current source). This scheme is only going to be stable if the phase-shift through the input stage is very low, and this is now actually less likely because there is less Miller feedback to reduce the input impedance of the VAS, therefore less of a pole-splitting effect, [6] and so there is more likely to be a significant pole at the output of the input stage. There is still some local feedback around the VAS, but what there is goes through the signal-dependent capacitance Cbc.

My experience with this configuration was that it was unstable, and any

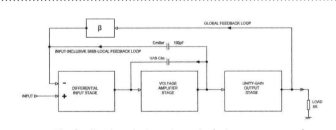

Figure 3   The feedback paths in an input-inclusive compensated amplifier.

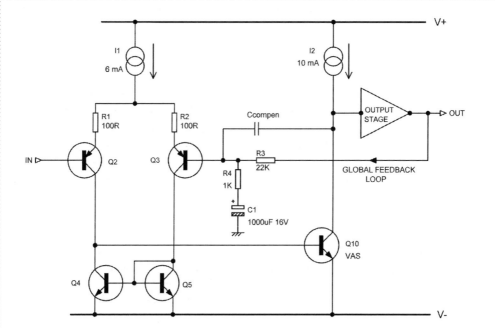

Figure 4   Attempt to implement input-inclusive compensation, with some typical circuit values shown.

supposed advantages it might have had were therefore irrelevant. I corresponded with JLH on this matter in 1994, hoping to find exactly how it was supposed to work, but no consensus on the matter could be reached.

A similar input-inclusive compensation configuration was put forward by Marshall Leach [7], the intention being not the reduction of distortion, but to decrease the liability of oscillation provoked by capacitive loading on the output, and so avoid the need for an output inductor.

Attempting to include the input stage in the inner loop to reduce its distortion seems to me to be missing the point. The fact of the matter is that the linearity of the input stage can be improved almost as much as you like, either by further increasing the tail current, and increasing the emitter degeneration resistors to maintain the transconductance, or by using a slightly more complex input stage. Any improvements in slew-rate that might be achieved would be of little importance, as obtaining a more than adequate slew-rate with the three-stage amplifier architecture is completely straightforward. [8]

If any stage needs more feedback around it is the output stage, as this will reduce its intractable

crossover distortion. I think that trying to create a semi-local loop around the input stage and VAS is heading off in wholly the wrong direction.

Blameless Power Amplifiers

Before we proceed to a form of output-inclusive compensation that does work, we need to take a look at the amplifier we are going to apply it to. Since we are studying only the compensation system rather than the whole amplifier, it makes sense to use a design which is both straightforward and known to have good performance with conventional compensation.

Some time ago I introduced the concept of the Blameless Audio Power Amplifier, [9] following an intensive study of the distortion mechanisms in power amplifiers. The Blameless title is intended to emphasise that although such amplifiers can give very low distortion figures, this is achieved more

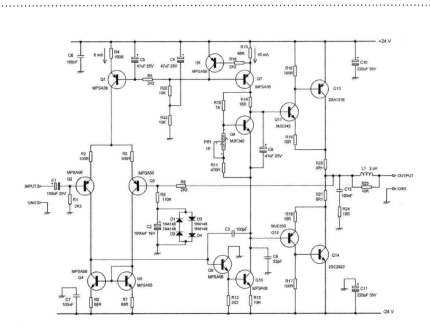

Figure 5   Schematic of a Blameless power amplifier with a CFP (complementary feedback pair) emitter-follower) output stage. Overload protection and clamp diodes are omitted for clarity.

by avoiding mistakes rather than by using radical and ingenious circuitry. The schematic of such an amplifier is shown in Figure 5; the following features make it a Blameless amplifier:

1.   The input differential pair Q2, Q3 is run at a higher current than usual, increasing the transcon-

ductance of the transistors. This is reduced to a suitable value for a stable amplifier by the emitter degeneration resistors R2, R3 which greatly linearise the transconductance of the input stage;

2. The collector currents of the input differential pair Q2, Q3 are forced into accurate equality by the current-mirror Q4, Q5, which is degenerated by R6, R7 for greater accuracy. This equality prevents the generation of second harmonic distortion in the input stage;

3. The VAS (Voltage Amplifier Stage) is linearised by adding the emitter-follower Q9, to increase the open-loop gain inside the local Miller feedback loop;

4. The output stage is in the CFP (Complementary Feedback Pair) configuration which has lower distortion than the EF (Emitter Follower) configuration in this situation;

5. The output stage emitter resistors are set to 0.1Ω, the lowest practicable value that permits acceptable quiescent stability. This reduces the inherent crossover distortion of the output stage with optimal biasing, and also minimises the extra distortion introduced if the output stage strays into Class AB operation due to over-biasing.

There are also topological requirements such as avoiding distortion being introduced by inductive coupling of half-wave currents from the supply connections.

While it is not directly related to distortion, note that the global feedback network R8, R9 has unusually low resistor values in order to improve the noise performance.

My intent was always that the Blameless amplifier was to be a sound and repeatable starting-point for innovative amplifier concepts. To be honest, it proved to be disconcertingly good, raising the awkward question as to why any further improvement might be necessary. Figure 6 shows that the distortion of a 22W/8Ω version is no more than 0.0001% at 400 Hz; it is unfortunately a good deal more at higher frequencies, and a long way short of perfection. Nonetheless, as hoped, this philosophy led to the Load-Invariant amplifier, [10] with unusually low levels of distortion into sub-8Ω loads, the Trimodal Amplifier [11] which was an ultra-low distortion Class-A

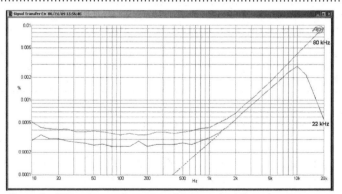

Figure 6   Distortion performance of the standard Blameless power amplifier at 22W/8Ω. THD at 10 kHz (80 kHz bandwidth) is 0.0042%. Extrapolating the 22 kHz trace to below the noise floor (dotted line) suggests that the distortion is no more than 0.0001% at 400 Hz.

amplifier that could be switched to Class-B, and was the basis on which the Crossover Displacement (Class-XD) principle was introduced [12]

Figure 6 shows the distortion performance of the design in Figure 5. The investigations reported here were done at the relatively low power of 22W into 8Ω, as this reduces the amount of damage that occurs if mistakes are made. With ±24 V rails as shown the maximum power output is about 30W into 8Ω,

The 80 kHz trace shows the THD rising at 6 dB/octave, because the feedback factor is falling at that rate due to the dominant-pole compensation. The 22 kHz trace attempts to show the very low distortion at low frequencies; with this bandwidth the noise floor is approximately 0.00025%. Above 10 kHz this trace falls at 18 dB/octave, the roll-off rate of the Audio Precision bandwidth-definition filter.

Stable output-inclusive compensation

Considering the problems described earlier, it is clear that trying to include the output stage in the VAS compensation loop over its full bandwidth looks impractical. What can be done, however, is to include it over the bandwidth that affects audio signals, but revert to purely local VAS compensation at higher frequencies where the extra phase-shifts of the output stage are evident. The method described here was suggested to me by the late Peter Baxandall, in a document I received in 1995, [13] commenting on some work I had done on the subject in 1994. He sent me six pages of theoretical analysis, but did not make it completely clear if he had personally evaluated it on a real amplifier. However he said that he had "devoted much thought and experiment to the problem" which implies that he had. Whether he invented the technique is not currently clear, but it is a characteristic Baxandall idea - simple but devastatingly effective. The general method was discussed in a DIYaudio forum [14] in March 2007; there appears to have been much simulation but no actual measuring.

The basic technique is shown in Figure 7. At low frequencies C1 and C2 have little effect, and the whole of the open-loop gain is available for negative feedback around the global feedback loop. As frequency increases, semi-local feedback through Ri and C1 begins to roll off the open-loop gain, but the output stage is still included in this semi-local loop. At higher frequencies still, where it is not feasible to include the output stage

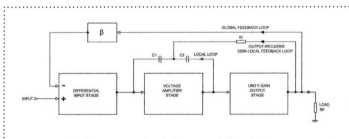

Figure 7   The basic principle of the Baxandall inclusive compensation technique.

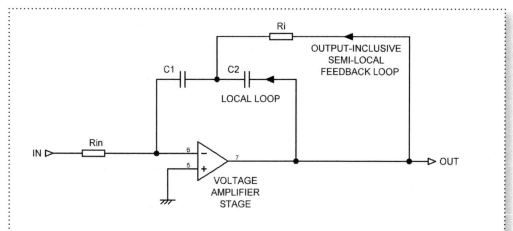

Figure 8   Conceptual diagram of the Baxandall output-inclusive compensation scheme for simulation.

in the semi-local loop, the impedance of C2 is becoming low compared to that of Ri, and the configuration smoothly changes again so that the local Miller loop gives dominant-pole compensation in the usual way. If the series combination of C1 and C2 gives the same capacitance as a normal dominant-pole Miller capacitor, then stability should be unchanged.

Figure 8 is a very much simplified version of a VAS compensated in this way, designed for simulation with a minimum of distracting complications. The current feed from the input stage is represented by Rin, which delivers a constant current as the opamp inverting input is at virtual ground. The "opamp" is in fact a VCVS (Voltage-Controlled Voltage Source) with a flat voltage gain of 10,000, once more to keep thing simple. Since there is no global feedback loop as in a complete amplifier, if the input is 1 Volt the output signal will be measured in kilo-Volts, at least at low frequencies; this is not exactly realistic but the magnitude does not alter the basic mechanism being studied.

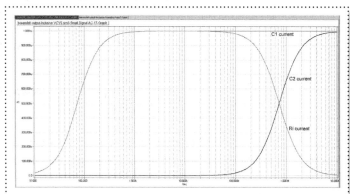

Figure 9   Showing how the local feedback loop (C2 current) takes over from the output-inclusive compensation (Ri current) as frequency increases.

Figure 9 shows how the feedback current through

C1 is sourced via Ri at low frequencies, and via C2 at high frequencies. The lower the value of Ri, the higher the frequency at which the transition between the two routes occurs. With C1 and C2 set at 220 pF, and Ri = 1K, this occurs at 723 kHz. At very high frequencies the effective Miller capacitance is 110 pF, a slight increase on the usual value of 100 pF; this simply because 220 pF capacitors are readily available.

At low frequencies the amount of semi-local feedback is controlled by the impedance of C1; at 220 pF it is more than twice the size of the usual 100 pF Miller capacitor, and so the reduced open-loop gain means the feedback factor available is actually 6.5 dB less than normal, at low frequencies only. This sounds like a bad thing, but the fact that the semi-local loop includes the output stage more than makes up for it. At 723 kHz the impedance of C1 has fallen to the point where it is equal to Ri. At 1.45 MHz the impedance of the series combination of C1 and C2 now reaches that of Ri, and the capacitors dominate, giving strictly local Miller compensation, with approximately the normal capacitance value (110pF).

Figure 10 attempts to illustrate this; it shows open-loop gain with the closed-loop gain (+27.2 dB) subtracted to give a plot of the feedback factor. Note that the kink in the plot only extends over an octave, and so in reality is a gentle transition between the two straight-line segments. This diagram is for Ri = 10KΩ, as this raises the gain plateau above the X-axis and makes thing a bit clearer.

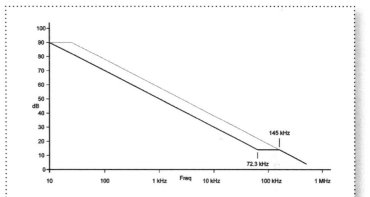

Figure 10 How the feedback factor varies with frequency. C1 = C2 = 220 pF, Ri = 10KΩ. The dotted line shows the feedback factor with normal Miller compensation with the capacitor equal to half C1, C2

The thin line shows normal Miller compensation; note that its feedback factor reaches a plateau around 20 Hz as this is the maximum gain of input stage and VAS combined, without compensation.

15

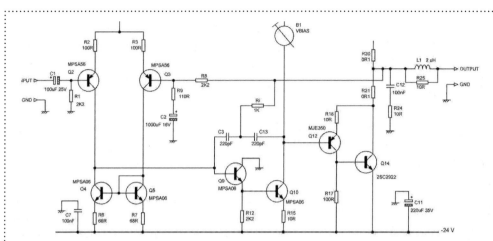

Figure 11 Practical implementation of output-inclusive compensation in the Blameless amplifier of Figure 5 (circuit simplified for clarity).

Figure 11 shows the practical implementation of the inclusive technique to the Blameless amplifier of Figure 5; the unaltered parts of the circuitry are omitted for greater clarity.

Figure 12 compares the distortion performance of the standard Blameless amplifier with the new output-inclusive version; there may be no official definition of "ultra-low distortion", but I reckon anything less than 0.001% at 10 kHz qualifies. The THD measurement system was an Audio Precision SYS-2702.

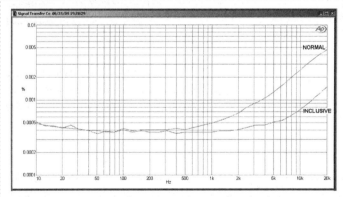

Figure 12 Distortion performance with normal and inclusive compensation, at 22W/8Ω. Inclusive compensation yields a THD + Noise figure of 0.00075% at 10 kHz, less than a third of the 0.0026% given by conventional compensation. Measurement bandwidth 80 kHz.

Figure 13 shows another view of the output-inclusive distortion performance. Because the THD levels are so low, noise is a significant part of the reading. Three measurement bandwidths are therefore shown. The 22 kHz and 30 kHz bandwidths eliminate most of the harmonics when the fundamental is 10

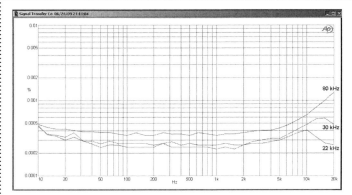

Figure 13 Bandwidths are 80 kHz, 30 kHz and 22 kHz.

kHz or above, so they do not give meaningful information in this region. It is however clear that a 10 kHz THD of 0.00074% is an overestimate, and 0.0005% is probably more accurate.

This form of compensation is very effective, but it is still necessary to optimise the quiescent bias of the Class-B output stage. This is easier to do without the distortion-suppression of output-inclusion, so you might consider adding a jumper so that the connection via Ri can be broken for bias setting.

As Figure 13 shows, this amplifier pretty much takes us to the limits of THD analysis; it is impossible to read any distortion at all below 2 kHz. The 22 kHz trace is only just above the AP distortion output; this varies slightly with output voltage but is below 0.00025% up to 10 kHz.

It is only right that I point out that a slight positional tweak of the output inductor was required to get the best THD figures. I attribute this to small amounts of uncorrected inductive distortion, where the half-wave currents couple into the input or feedback paths; it is an insidious cause of non-linearity. [15] The amplifier I used for the tests was thought to be free from this, but the reduced output stage distortion appears to have exposed some remaining vestiges of it. At 10 kHz the inductive coupling will be ten times greater than at 1 kHz, but in a conventional Blameless amplifier this effect is normally masked by crossover products.

The information here is not claimed to be a fully worked-out design such as you might put into quantity production. As with any unconventional compensation system, it would be highly desirable to check the HF stability at higher powers, with 4Ω loads and below, and with highly reactive loads.

Practical experimentation
One of the great advantages of this approach is that it can be added to an existing amplifier for the cost of a few pence. Possibly one of the best bargains in audio! This does not mean, of course that it can be applied to any old amplifier; it can only work in a three-stage architecture, and the amplifier needs to be Blameless to begin with to get the full benefit.

Another advantage of this arrangement is that if you decide that inclusive compensation is not for you, the amplifier can be instantly converted back to standard Miller compensation by breaking the connection to Ri. There is in fact a continuum between conventional and output-inclusive compensation. As the value of Ri increases, the local/semi-local transition occurs at lower and lower frequencies.

The quickest and most effective way to experiment with this form of compensation is to start with an amplifier that is known to be Blameless. A very suitable design is the Load-Invariant amplifier produced by The Signal Transfer Company; [16] this has all the circuit features described above for Blameless performance, and the PCB layout is carefully optimised to eliminate inductive distortion. I need to declare an interest here; I am, with my colleague Gareth Connor, the technical management of The Signal Transfer Company.

Conclusion

It is instructive to compare this method with other ways of achieving ultra-low distortion. The Halcro power amplifier is noted for its low distortion, achieved by applying error-correction techniques to the FET output stage; however what appears to be a basic version, as disclosed in a patent, [17] uses 31 transistors. Giovanni Stochino (for whose abilities I have great respect) has also used error-correction in a most ingenious form, [18] but it requires two separate amplifiers to implement; the main amplifier uses 20 transistors and the auxiliary correction amplifier has 17, totalling 37. The output-inclusive compensation amplifier uses only 13 transistors, and gives extraordinary results for such simple circuitry.

Is this the end of history for power amplifiers? Not if I have anything to do with it.

REFERENCES

[1] Douglas Self, Audio Power Amplifier Design Handbook , 5th edition, Newnes, p198   (two-pole compensation)

[2] John M Miller, Dependence of the input impedance of a three-electrode vacuum tube upon the load in the plate circuit, Scientific Papers of the Bureau of Standards, 15(351):367{385, 1920

[3] Douglas Self, Audio Power Amplifier Design Handbook, 5th edition, Newnes, p119   (VAS enhancements)

[4] Widlar, R A, Monolithic Power Op-Amp, IEEE J Solid-State Circuits, Vol 23, No 2, April 1988

[5] John Linsley-Hood, Solid-State Audio Power, Electronics World +Wireless World, Nov 1989, p1047

[6] Douglas Self, Audio Power Amplifier Design Handbook , 5th edition, Newnes, p198   (pole-splitting)

[7] Marshall Leach, Feedforward Compensation of the Amplifier Output Stage for Improved Stability

with Capacitive Loads, IEE Trans on Consumer Electronics, Vol 34, No 2, May 1988

(So far as I can see, this is a misuse of the word "feedforward". What is actually described is semi-local feedback around the input stage, so that the output stage actually sees less HF feedback. The aim is to avoid using an output inductor; I feel this is a mistake. The amplifier is divided into only two stages for analysis)

[8] Douglas Self, Audio Power Amplifier Design Handbook, 5th edition, Newnes, p255 (slew-rates)

[9] Douglas Self, Distortion In Power Amplifiers: Parts 1 to 8, Electronics World Aug 1993 to Mar 1994

[10] Douglas Self, Load-invariant audio power, Electronics World Jan 1997 p16

[11] Douglas Self, Trimodal audio power, Part 1: Electronics World June 1995 p462, Part 2: Electronics World July 1995 p584

[12] Douglas Self, Class XD: a New Power Amplifier, Electronics World Nov 2006 p20

[13] Peter Baxandall, Private communication, 1995

[14]http://www.diyaudio.com/forums/solid-state/94676-bob-cordell-interview-negative-feedback-49.html (Pages 49 - 67)

[15] Douglas Self, Audio Power Amplifier Design Handbook, 5th edition, Newnes, p198 (inductive distortion)

[16] The Signal Transfer Company: http://www.signaltransfer.freeuk.com/

[17] Bruce Halcro Candy, US patent No. 5,892,398. (1999) Figure 2

[18] Giovanni Stochino, Audio Design Leaps Forward? Electronics World, Oct 1994, p. 818

The Arch Nemesis

Nelson Pass

Introduction

A poster of Einstein once said, "Things should be made as simple as possible, but no simpler". This can apply to audio amplifiers, but if they are evaluated subjectively, the simplicity thing can get a little of hand. Of itself, minimalism exerts a strong aesthetic attraction, and there is a reasonable belief that fewer components in the signal path allows more information to get through with less coloration.

If like me you are interested in understanding of how we hear distortions with our brains (instead of our meters), you might appreciate that simple circuits help isolate these phenomena. I listen to all sorts of flawed circuits because I enjoy hearing the differences, and it helps to train my ears. In this regard, reducing the number and types of flaws makes it easier to tweak a single parameter and hear the difference. I think it's also true that simple distortions are often more forgivable in a listening situation and create less fatigue.

The Nemesis

In 1985 Jean Hiraga wrote an article in two parts presenting, among other things, a design for a very simple MOSFET amplifier called the Nemesis. Subtitled "An Homage to the WE 25 B", the piece celebrated classic simplicity in amplifier design, specifically a Western Electric amplifier that used a single gain triode driven by an input transformer and driving an output transformer, as shown (simplified) in Figure 1.

Hiraga also discussed a 1982 amplifier apparently done as an application note for Siliconix using the VN64GA N channel power MOSFET

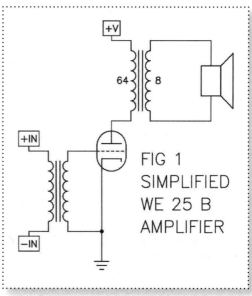

FIG 1
SIMPLIFIED
WE 25 B
AMPLIFIER

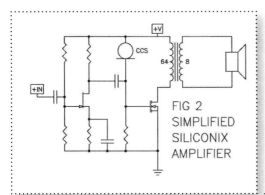

FIG 2
SIMPLIFIED
SILICONIX
AMPLIFIER

driven by a J106 JFET shown in Figure 2. He went on to simplify this circuit by eliminating the input JFET, driving the Gate of the power MOSFET directly (Figure 3).

A second version had an interesting connection from the Source of the transistor to the secondary winding, providing both feedback for the transistor and some "auto-former support" to the output secondary (Figure 4).

But Part 2 of L'amplificateur Némésis showed the final schematic where the transistor feedback connection to the secondary winding was dropped, the simplified circuit reverting back to Figure 3. It appears that he was more interested in the sound without the feedback, even though the measured performance suffered.

In 1994 I played with similar concepts in the Zen Amplifier (Figure 5) and followed up with a series of variations on the theme which explored single-transistor designs, some with feedback and some without, but none of them employ-

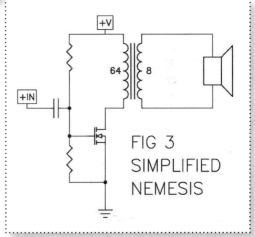

FIG 3
SIMPLIFIED
NEMESIS

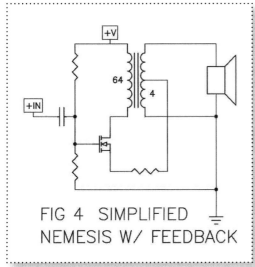

FIG 4 SIMPLIFIED
NEMESIS W/ FEEDBACK

ing output transformers. These articles can be downloaded from www.passdiy.com. At the time I was solely interested in the performance obtainable from single Class A gain stages alone and didn't want to also consider the additional distortions of passive components (including transformers) in the signal path.

But there was another reason for not using transformers in the Zen amplifiers - the power MOSFETs involved are already pretty happy at the voltages and currents needed by loudspeakers. Tube circuits operate at higher voltages and lower currents by a factor of about 10,

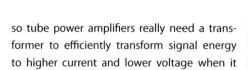

so tube power amplifiers really need a transformer to efficiently transform signal energy to higher current and lower voltage when it comes to driving 8 ohms.

The Zen amplifier philosophy ("What is the sound of one transistor clapping?") calls for a minimum of parts. A component has to be needed to be included, but if you alter the need criterion from "measuring better" to "sounding better" then a potentially different perspective opens up.

Brief Digression...

It is a common belief among audiophiles that measurements don't correlate all that well with

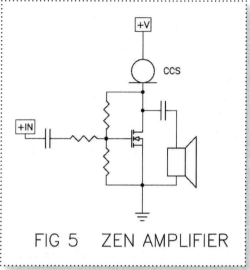

FIG 5   ZEN AMPLIFIER

subjective experience. This is not very surprising – the ear/brain is immensely complicated, and there are many experiments to demonstrate that our understanding of hearing is not much better than our understanding of consciousness, which is not good. Simply the fact that different cultures and individuals hear known "audio illusions" differently gives us a clue while making the problem seem more intractable. I don't expect it to be well understood in my lifetime.

Some "objectivists" think that audiophile subjectivism is delusional, and they are often right, but that doesn't mean that people hear the same way as test equipment. In the first half of the 20th century, there was a reasonably clear association between measured performance and perceived performance, but probably this was due to the rather high distortion of early equipment, where 1% distortion was considered quite good. These days it's common to see amplifiers measuring .001% or even less, but the audio marketplace doesn't seem to particularly reward such an achievement.

Back to our Program...

One of the charms of simple circuits is that they have a better correlation between objective (measured) and subjective (heard) performance. It seems that a simple circuit that measures good is more likely to sound good than a complicated circuit that measures good. Moreover, It appears that simple amplifiers like Nemesis and the Zen make it easier to hear differences between single components and compare these subjective differences to measurements.

So Why an Output Transformer?

All components have distortion. We can rank them pretty easily based on simple measurements like THD (total harmonic distortion) and variations they cause in frequency response. Wire and resistors

are at the top of the list because as a rule they measure quite low. Next are the capacitors, which give us low but easily measured distortions. At the bottom are active gain devices such as tubes and transistors. And transformers.

Signal transformers don't tend to get a lot of respect from objectivist solid state guys due to bandwidth and distortion issues. You can build a good transistor amplifier without them, and so most do. But Hiraga was (is) not a fool, and in addition there is a small audio cult that likes transformers, even when they aren't essential. They use them for output coupling, input coupling, volume controls and passive crossovers. These are often the same people who disdain capacitors with nearly the same emotion they reserve for MP3 compression.

What's wrong with these people and what is it with transformers?

Jan Didden's Nemesis

Who knows what's wrong with Jan? Whatever it is, apparently he addressed it by building himself a copy of the Nemesis. I heard about it because he also seems to have wanted to make more of them and talked to Jack Elliano at www.electra-print.com about getting some transformers made. Ultimately he decided that shipping to Europe was too expensive - did anyone on the forum at www. diyaudio.com want to take up the project? ...That would be me.

I have spent quality time with coupling transformers before, but I had never really warmed up to them, possibly because I had not yet reached the 10,000 hour level of listening required to achieve audiophile expertise. In any case, a couple of years ago I began experimenting with transformers to solve some problems in a couple of future Zen amplifier projects, and got some fairly good results (good enough for Zens, anyway). Having worked out some circuits, I acquired an assortment of transformers and began evaluating their performance with an eye toward picking the best one. They represented a wide range of cost and materials, and some clearly measured better than others, but when I listened to them I found myself drawn to the sound of one that didn't measure so well.

The dissonance that measures bad/sounds good created called for an unbiased test. So I built two identical amplifiers except for transformers – the very expensive one which measured best, and the unpretentious one that didn't measure so well. I packed them off for a reliable blind test with Joe Sammut, who has 10,000 hours more listening time than me.

"This one is really musical, and that one is not very good."

Well, that's another data point – a transformer that measures better loses to one that does not. Perhaps if my French was any good, Hiraga would have explained it to me long ago.

The Arch

Jack sent me a nice pair of transformers, and I set about making a simple recreation of the Nemesis but with variable values for supply voltage, input DC bias, Source resistance, and resistance across both the primary and secondary coils of the transformer, as shown in Figure 6.

As the schematic reveals, there is a lot of opportunity to play around. The input bias voltage ranges over +/- 10V DC. For enhancement-mode MOSFETs and JFETs, it may require a positive voltage as high as 8 volts or so. For depletion-mode devices the bias voltage will range from as low as -5 volts to as high as +2 volt. The 10 KOhm resistor between the BIAS voltage and the Gate of the transistor is arbitrary. I used this value because Hiraga did, but you can consider values as high as 100 Kohm for MOSFETs and depletion-mode JFETs. If you see more than 100 mV DC across it with an enhancement-mode JFET, then you might want to reduce the value, but it's not a big deal.

Not shown, but you may want to consider a 50 to 100 ohm resistor in series with the Gate of the transistor. This is customary, but I did not experience issues in this amplifier without it. If you have issues with high frequency oscillation, you will want to insert one. Of course if you are using MOSFETs, you need to avoid zapping the Gate with a static shock. Elementary caution is usually more than adequate.

Typically the main power supply will range from about 30 to 40 volts, but you can go lower or higher if you want, within the dissipation limits of the transistors. As it is, I ended up dissipating about 40 watts in a single transistor with 35 volts, which is pretty close to the limit. Since 30 volt supplies are common, you should feel free to use that value if it's convenient.

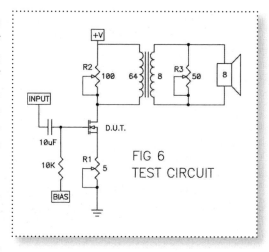

FIG 6
TEST CIRCUIT

First, I wanted to explore the limitations of the transformer. It is a single-ended design with a 64 ohm primary and 8 ohm secondary, which is about a 2.8 to 1 turns ratio. The maximum primary DC current is rated at 1.3 amps. Electra Print's specification for bandwidth and distortion was taken with a 25 ohm source impedance driving the primary.

The bias current and source impedance are important factors in this circuit as they have a strong influence on the distortion and frequency response, particularly at low frequencies. If the bias current is too high, the transformer saturates at low frequencies and the distortion goes up and the frequency response suffers, as seen in Figure 7. Here we see an example of this circuit where only the bias is

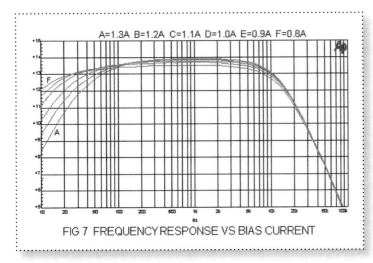

FIG 7 FREQUENCY RESPONSE VS BIAS CURRENT

varied, and where higher current through the primary creates greater roll-off at the bottom end.

Figure 8 shows an example of distortion as a function of bias current, and we see that for circuits of this type lower bias improves the bottom end, but higher bias improves the midrange. Higher bias improves the performance of the gain device, and incidentally allows for greater power. You can appreciate that performance trade-offs will be involved.

The source impedance of the circuit driving the primary of the transformer has a similar effect. This transformer was designed around a 25 ohm source. A typical MOSFET operated single-ended Class A as in Figure 6 has an intrinsic output (Drain) impedance of a couple hundred ohms or so. In Figure 6 you will see a variable resistor R2 which can be used to adjust the source impedance seen by the transformer primary. In an example test with a 1 amp bias current we see that the low frequency roll-off (-3dB) is at 40 Hz. With R2 at 75 ohms it's 25 Hz, and with 36 ohms it's 18 Hz. Distortion degradation with higher source impedance is comparable to the example of Figure 8.

As with bias current, adjusting source impedance gives us the opportunity to examine potential performance compromises. Lowering the source impedance via R2 improves transformer performance, and lowers the output impedance of the amplifier as a whole (more damping factor for the loudspeaker), but it loads the gain device, making it work harder to deliver the voltage we want and creating more distortion as a result. The reason that R1, R2, and R3 are variable in this circuit is to afford the opportunity to adjust and

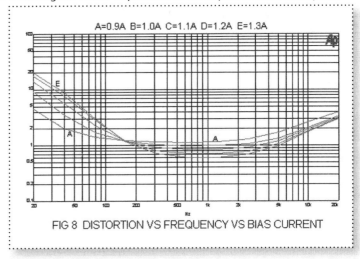

FIG 8 DISTORTION VS FREQUENCY VS BIAS CURRENT

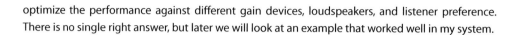
optimize the performance against different gain devices, loudspeakers, and listener preference. There is no single right answer, but later we will look at an example that worked well in my system.

For gain devices I had the old standby IRFP240 N channel enhancement-mode MOSFET, plus Ixys' IXTH6N50D2 and IXTH20N50D depletion-mode MOSFETs left over from the De-Lite amplifier (www. diyaudio.com).

In addition to MOSFETs, I had three examples of SemiSouth power JFETs, the enhancement-mode SJEP120R100 and SJEP170R550, and the depletion-mode SJDP120R085. This last part almost didn't make it into this project, as I was not prepared to talk about it until it was publicly disclosed. The SemiSouth parts are made of Silicon Carbide (SiC) and while designed for fast high efficiency switching, they turn out to have superior linearity, resulting in lower distortion.

As a first step, I decided to create an apples-to-apples comparison of the performance of the gain devices. Using the circuit of Figure 6, I set the supply voltage at 32 volts, R1 at 1 ohm, R2 at 36 ohms, and R3 open. The voltage at the Bias pin was varied for each device to give a 1.2 amp bias current.

Each device was measured for response, distortion vs output power, and distortion vs frequency. These figures were taken into an 8 ohm load, and the response and distortion vs frequency were taken at 1 watt and with a 25 ohm and a 600 ohm source impedance from the input signal generator.

Figure 9 shows a table summarizing the results. Not shown is the frequency response, which was consistently -3dB at 25 Hz and 25 KHz with both the 25 and 600 ohm source impedance. All the parts had essentially the same 1 watt distortion at 1 KHz and 20 KHz with a 25 ohm source imped-

	Gain DB	THD 1 W	THD 4 W	THD 20 KHZ @ 600 OHM
SJDP120R085	15.5	0.6%	1.7%	3.0%
SJEP120R100	16.4	0.9%	2.4%	5.0%
IRFP240	15.5	1.5%	4.0%	6.0%
SJDP170R550	15.9	1.5%	3.0%	2.0%
IXTH6N50D2	14.7	1.8%	4.0%	10.0%
IXTH20N50D	14.6	2.6%	6.0%	5.0%

Figure 9

ance, but all had more distortion at 20 KHz with 600 ohms, and so I include that data.

All of these parts work well enough to use, and we note that the ubiquitous and cheap IRFP240 is no slouch; however two parts stand out. The SJDP120R085 JFET has the lowest distortion at all power levels and frequencies, except at 20 KHz (600 ohm source), where the SJEP170R550 beats it due to its low input capacitance.

I don't have the time and space to examine all the permutations for different parts values with the different devices, but we will look at the

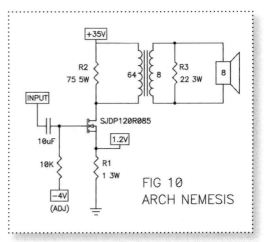

most interesting example, the SJDP120R085. Playing around semi-randomly with the adjustable values and test equipment, I settled on a set which looked like a good compromise, and I present it here in Figure 10.

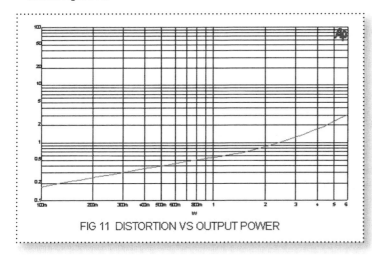

FIG 11 DISTORTION VS OUTPUT POWER

Figure 11 shows the distortion vs amplitude for this version of the amplifier.

Figure 12 shows the distortion vs frequency at 1 watt.

I adjusted R3 to give the amplifier a damping factor close to 1, and while this could have been done at R2, it's nice to have a little resistance on the output.

Comparing Jan's Nemesis figures to the Arch, I note that they both would be considered about 6 watt amplifiers, with Jan's having 3 or 4 dB more gain. The Arch achieves about one third the distortion in the midband at 1 watt, and slightly better numbers than Jan's at 6 watts. The top end of Jan's transformer has similar distortion figures, but has approximately an octave more bandwidth. At the bottom, Jan's version does not make it quite as low, but has a bit less distortion.

Offhand I think Jan's transformer is a little better, and I'm willing to bet it was a lot more expensive than Electra Print's. It looks as though the SemiSouth power JFET was better than Jan's transistor

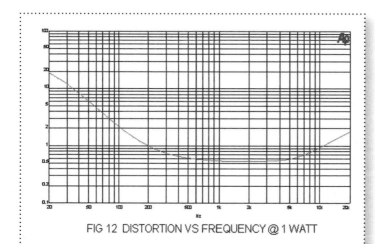

FIG 12 DISTORTION VS FREQUENCY @ 1 WATT

(which was a lateral Exicon ECF10N16 – ed.), and I know it cost a lot more. All in all, I would expect a strong family resemblance between the two amplifiers.

So How Does It Sound?

It's not easy carrying bread-boarded amplifier channels over to the HiFi system. I have learned to glue the parts down so they don't fall off the plywood. I recommend a hot glue gun, as no one wants to wait while silicone sets up. I also have learned to clean up my soldering and tie down my wires, lest they get snagged and short. No clip leads, and commercial regulated power supplies.

Having accomplished the move, I hooked it up to my bi-amped open baffles driving either a pair of Lowther PM5A's or Feastrex D9nf from 150 Hz on up. Both loads are 16 ohm with efficiencies of 96 and 92 dB respectively. This represents a fortuitous setup for this amplifier, as it doesn't have a lot of power or damping factor, doesn't ask for much on the bottom end, and doesn't have more bandwidth than the amplifier.

My comparison amplifier was an F2J (an F2 using the SJEP120R100) which has comparable gain, mid-band distortion, second harmonic characteristic and was also set for a damping factor of 1. It was a good match, particularly as the F2J has a very much wider bandwidth which I would expect to highlight the top end differences with the transformer.

You may have guessed by now that I like simple music. Single guitar and vocals, low key jazz trios and quartets from the 50's and early 60's, Ella Fitzgerald, Sarah Vaughan, Nina Simone, contemporary lounge, and post-modern classical. These types bring out the best in full range drivers and little amplifiers. I have a very scientific procedure for evaluating the sound – I sit and listen to whatever I like for as long as I want. I sincerely try to get the best out of each thing tested and usually it takes a couple days of part time effort, and inevitably an impression emerges into words. Mostly I find myself looking for sound that relaxes me while still presenting a lot of information.

To cut to the result, this amplifier does just about that. I believe I can hear the limited bandwidth, but it doesn't matter a lot. The sound is a little more tube-y than the F2J, but that makes sense – there's a transformer in there. All told it's very musical and toe-tapping.

One of the interesting things about this amplifier as compared to the F2J is that it seems as if the limited high frequency bandwidth tends to focus your attention on the midrange, which as Paul Klipsch said, is where we live. It's as if the lesser distraction brings midrange into sharper relief, paradoxically revealing more detail. I recall my impressions of the single-ended tube amplifiers I've had in the system, and it seems to me that the Arch Nemesis manages the lower distortion of those that used feedback and some of the less effable character of the non-feedback types. Toward that end, I think it offers a bit of both. I think I can safely recommend this for full range driver and SET enthusiasts, particularly if you like to experiment with tweaking the sound. It's not a triode, but it gets close. If your preamp has an output impedance of 1Kohm or greater, then you should consider using the SRDP1700R550 for its low input capacitance or a buffer, such as the B1 circuit. (www.firstwatt.com)

Construction Notes

I have tried to structure these schematics in such a way as to encourage experimentation should you decide to build this amplifier. None of the values or parts are cast in cement, and you should feel free to play with different parts and values. In addition to the usual cautions about high voltage, I will add that you want to avoid running too much current through the transistor and you want to make sure the transistor gets enough heat sink – probably about 0.5 deg C per watt per channel.

It is my understanding that Jack at www.electra-print.com will be offering the transformers for sale at reasonable prices. They are recommended, but of course there may be others available. As long as they resemble the specs for the one used here, then they will at least work.

I used commercially available regulated variable power supplies. Some of these are available at modest cost. At www.mpja.com there is a Mastech HY3003F-3 which offers a dual 30V @ 3A variable with an isolated 5 volt fixed which can be used for the bias voltage with depletion-mode devices. These have the convenient features of limiting and readouts of voltage and current for each channel, but be aware that they fan cool if they get hot.

Figure 13 shows how you might decide to configure the bias for enhancement-mode devices.

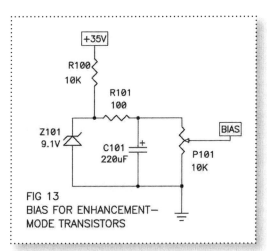

FIG 13
BIAS FOR ENHANCEMENT—
MODE TRANSISTORS

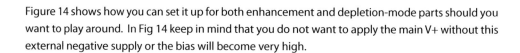

Figure 14 shows how you can set it up for both enhancement and depletion-mode parts should you want to play around. In Fig 14 keep in mind that you do not want to apply the main V+ without this external negative supply or the bias will become very high.

You can build your own supply as you like, for instance those that I published for the Zen variations on www.passdiy.com. A regulated supply is described in The Zen Variations - Part 3. A tip: if you want to place the mains transformer on the same chassis as the amplifier, you must avoid hum coupling from mains to output transformer. Play

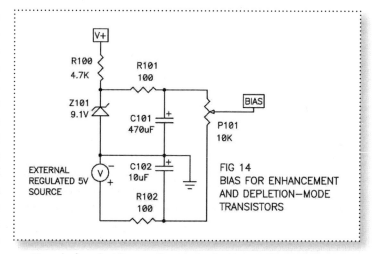

with the relative transformer positions before deciding on the mechanical layout. Keeping them as far apart as possible and at right angles to each other should help. And, of course you can always visit the happy helpers at the Pass Forum at www.diyaudio.com for advice and camaraderie.

Down the Rabbit Hole - Adventures in the Land of Phonostages

Joachim Gerhard

My obsession with phonostages goes back a long way. In the early 80s I designed a Pre-Pre (today the more common term is Head Amp). That was also the first product under the Audio Physic label and not a loudspeaker as many may think. It was a parallel symmetric common base stage without global feedback and floating battery supply. I got the circuit idea from Wireless World and had at that time no clue that it was based on a circuit by Prof. Marshall Leach that preceded that design by several years. Today that circuit has survived in the form of the LC Audio MC Pre with a twist of a „light" supply (Fig1).

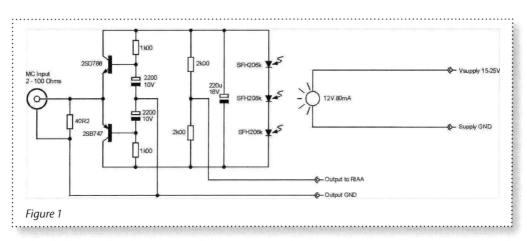

Figure 1

The following 20 years saw the growth of Audio Physic and I was fully occupied with developing loudspeakers and more and more management duties. My love affair with phonostages and particular the problem of amplifying and equalizing tiny signals did not stop though and instead of developing something myself I bought every phonostage I found interesting, dissected it and sometimes modified it. So over the years I amassed an amazing collection of famous and not so well known stages including designs by Mike Bladelius, Prof. Hawksford, Dr. Bews, Reiner Reddemann, Jonathan Carr, Stephan Horwege, LC Audio and Jürgen Ultee. Usually I opened up the case, analyzed

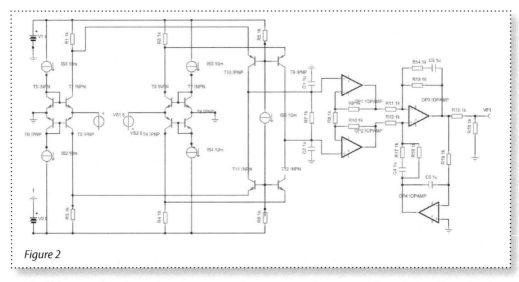

Figure 2

the circuit and started conversation with the designer. Over the years I collected a lot of knowledge and one day this information container simply overflowed and a plethora of circuits materialized in my head. An urge to put that information into working circuits followed and I started my adventure with opening the MPP thread on diyaudio a year ago.

The first design I posted was a balanced MC stage that also had parallel symmetric common base stages. In some circles this is known as zero Ohm input but being an open loop design it actually has around 10 to 20 Ohm input impedance, depending on the transistors and the idle current (Fig.2). I had used the MAT02 and MAT03 transistors in a desire to use „the best" but I soon learned that they are not particular symmetric concerning matching of P and N variety and also have been super-seded by other BJT´s that have even lower noise, better matching and higher and more linear Hfe. (Fig 3 lists some candidates). The critical parameter for low voltage noise is Rbb, the base spread-ing resistance. Low voltage noise is important when designing an input stage for MC cartridges because they usually have a low internal impedance. It can be as low as 2 Ohms and up to 50 Ohms but a relatively high impedance plus a low output voltage is not a good start for a low noise result because the Johnson noise of the cartridge adds to the noise of the phonostage. Ideally an MC cartridge should have the lowest possible impedance and the highest possible output voltage so a cartridge with low impedance and very low output voltage is a challenge too be-cause the phonostage has to make up the gain and loses Signal-Noise Ratio in the process. (Fig.4 lists some common and not so common

Figure 3
List of ultra low noise bipolar transistors

2SA1316 (Rbb = 2 Ohm)	2SC3329 (Rbb = 2 Ohm)	Toshiba
2SB737 (Rbb = 2 Ohm)	2SD786 (Rbb = 4 Ohm)	Rohm
2SA1085 (Rbb = 1.7 Ohm)	2SC2547 (Rbb = 1.7 Ohm)	Hitachi – Renaisas

These transistors are designed for Audio and also have high and linear Hfe

Figure 4	List of MC Cartridges : Ohmic impedance and output voltage at 5cm/sec @1kHz	
Audio Note IO Ltd.	1 Ohm	0.04mV
My Sonic Lab Hyper Eminent	1.8 Ohm	0.5mV
Ikeda 9REX	3 Ohm	0.18mV
Ortofon MC A90	4 Ohm	0.27mV
Ortofon MC20 Supreme	5 Ohm	0.5mV
Lyra Olympos	5.5 Ohm	0.3mV
Lyra Titan i	5.5 Ohm	0.5mV
Ortofon SPU Classic GM Mk2	6 Ohm	0.2mV
Van den Hul Colibri XCP	21 Ohm	0.4mV
EMT XSD15	24 Ohm	1.05mV
Dynavector DV DRT XV1t	24 Ohm	0.35mV
Clearaudio Stradivari MC	35 Ohm	0.8mV
Denon DL103	40 Ohm	0.3mV
Benz Micro H2	90 Ohm	2.5mV

examples). You can draw your own conclusions but I find the Hyper Eminent outstanding in relation to impedance – output voltage ratio. There must be some extremely strong magnets at work here. Stig Bjørge from Lyra told me that they have access to very strong magnets too, but the real secret of the Eminent cartridge is a core material with extremely high permeability, and they will not tell what it is. In praxis a lot of issues play a role for the potential sound, not the least the choice of diamond cutting and grinding, the suspension and mechanical design in terms of rigidity and damping.

I am using Lyra cartridges most of the time and get very quiet results with the circuit that will evolve here even on my 100dB sensitive MPL loudspeaker system. The hum of the power amps I tried (especially the Tube variety) is most of the time higher than hum and hiss from this Phonostage.

So back to phonostages.
After I had designed a lot of Low Z BJT stages it occurred to me that DC coupling is not easy to achieve if you follow that road. A DC current through the cartridge is to be avoided because it generates 2nd harmonic distortion and in a worst case scenario can destroy the delicate wires. Vbe is a very weak function of idle current so making a DC coupled input and output comes with considerable complexity. DC canceling in the input stage is possible but comes with a noise penalty and additional circuitry that is against Occam's Razor, the notion that the simplest solution that does the job is often also the best. Words like loss of transparency come to mind. I did design the Ultra Low Noise

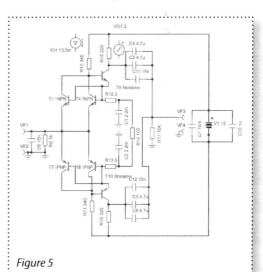

Figure 5

33

ROSI that avoids DC in the cartridge with a floating supply. This is an evolution of a circuit that was presented by Rositter in AudioXpress. Unfortunately it needs a big electrolytic cap to ground and is not DC coupled at the output (Fig.5). Nevertheless circuits like that can sound excellent if some care is taken in component choice. They major in drama and slam, possibly because the movement of the stylus is somewhat damped by the low input impedance.

The next logical step was to try common emitter or emitter followers, with and without bias cancellation. Both circuits have the advantage of low DC at the input because only little current is flowing in the base. Take that idea further and FETs are unavoidable. Compared to BJTs they have nearly zero gate current so DC coupling is a breeze. Their main disadvantage is a quite high 1/f noise corner frequency so they are more noisy in the lower ranges of the frequency spectrum. In praxis I found that less of a problem because the human ear is not very sensitive at lower frequencies and the rise in noise is very shallow in modern, optimized devices. Higher up the frequency range the noise from a FET has a more soft „benign" tone so low noise designs with FETs can be done.

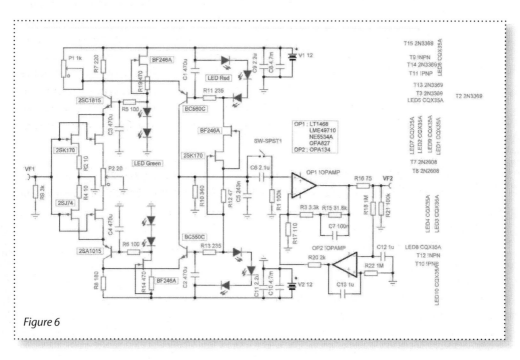

Figure 6

The first design I did was an instant hit and is now known as the High Z MPP. Several people build it with success. It is simple, high performance and elegant. It is a folded cascode open loop design. Noise can be brought down to 0.3nV/RtHz if 8 FETs are used. Distortion is -100dB at 2 V output theoretically (I measured -90dB in praxis) and frequency response is wide (-3dB over 5MHz; Fig6).

Whatever I tried, I could not improve distortion with an open loop design, so this is at least for me the limit what can be done with open loop designs. That notion was confirmed when I saw the measurement of „The Grail", the new Van den Hul Phonostage. It uses an open loop input circuit and was designed by Jürgen Ultee, an expert on measurements.

So when you are a fan of open loop design (and there are many), look no further.
When you belong to the chosen few that can hear distortion less then -90dB and you are not afraid of feedback (highly optimized, that is) I have a gift for you:

The FPS - Flexible Phono Stage.

So NFB (negative feedback) was on the agenda. Strictly speaking an open loop design also needs feedback to be stable, in this case a resistive element that does „current feedback". The difference to NFB is that negative feedback returns a part of the output signal to the input for error cancelation. Some designers I talked to and that I respect told me that part of the input signal could be lost in the process especially considering the tiny signals involved and that the distortion gets lower but more complex in spread. The result would then be a loss of acoustic information leading to a flat „papery" soundstage and a loss of micro information making the sound sterile, analytic and hard. My head was humming and I calculated the numbers even while in the car and in my dreams. Considering an output of 0.5mV at 5m/sec and 20dB headroom there still can be -60dB „bottom room" so the lowest encoded signal would be 0.5uV. Considering 40dB of feedback we have a correction signal of 5nV, not much. The only way out I found was finding a topology that has superb low level resolution, building something and then listen if I could detect a loss of information compared to the open loop circuits I had already built.

I have worked in the High Tech industry from 1985 to 1989 and knew that there is a topology that is optimized to amplify tiny signals: The Instrumentation Amplifier or INA in short. It is for example used to detect brain waves. In its basic form it consists of 3 operational amplifiers that are connected in a certain way. One of the most recent integrated solutions is the INA163 from Burr-Brown (Fig 7). If you want to know more about INAs you can download „ A Designers Guide to Instrumentation Amplifiers, 3rd Edition" from www.analog.com . On the first page you can find the phrase: „An Instrumentation Amplifiers is a device that amplifies the DIFFERENCE between two input signal voltages while rejecting any signals that are common for both inputs. The in-amp therefore, provides the very

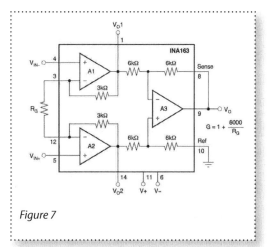

Figure 7

important function of extracting small signals from transducers and other signal sources", and that is exactly what we want. The first circuit I tested used the INA163 at the input. It certainly sounded great with a lot of dynamic slam. I could not detect a loss of information so this was a very encouraging result. One of the disadvantages of an integrated INA is that the internal feedback resistors are fixed. In the case of the INA163 they are 3kOhm. The gain equation of an INA is Gain = ((2 x R1) / R2) +1 so for a gain of 36dB R2 ends up to be 80 Ohms. Unfortunately that resistor is in series with the signal so its Johnson noise adds to the noise of the active stages in the INA. In this particular case

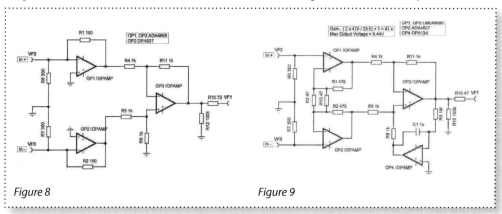

Figure 8   Figure 9

the ca. 1.3nV/rtHz of the 80 Ohm resistor adds to the 1nV/rtHz noise from the active stages in the INA163. One solution is to raise the gain and then put a resistive divider at the output. That would make R2 small but feedback is less so distortion goes up and bandwidth goes down, quite a waste of quality. Another way is to parallel several devices to lower the voltage noise. This arrangement is called a Faulkner Circuit. A friend of mine tried that with the INA163 but he could not get that circuit stable.

To circumvent this situation I developed circuits that use 3 separate OP amps, so I have a wider choice for R1. I designed a Low Z (Fig.8) and High Z (Fig.9) version and they worked beautifully. Again the Low Z version had a lot of slam and raw dynamics and the High Z version had a somewhat smoother sound with elegant rendering of tonal shades. From the technical side the Low Z version has the benefit that cartridge impedance is part of the feedback structure so a cartridge with low impedance that has usually a low output voltage gets more amplification than one that has high impedance but more voltage output. Cartridge and INA form a balanced shunt feedback structure and it could be argued that this reduces distortion. The current feedback architecture of the INA163 avoids the pitfall of lower bandwith with high gain but still distortion lowering feedback is reduced with higher gains. Another advantage is that there is no other potentially noisy resistor in series with the cartridge. The High Z version on the other hand has the advantage that gain can easily be changed with only one resistor and cartridge loading can be varied too over a huge range. A lot of

people believe that cartridge loading effects the sound much, not only the value of the resistor but also the type, so here you have a huge playing field. At first I designed servos into both circuits but I found it unnecessary because one of the advantages of an INA is low DC offset. In both circuits I tried, offset was not much more than 2mV. If that is too high for your system use the servo, for example with an OPA827 that has 0.1mV offset. The popular OPA134 has around 1mV DC in this application and would not bring a substantial benefit unless you are concerned about the sound. Actually a DC servo in an INA can potentially sound better because the output stage is lifted from ground and connected to a virtual ground provided by the servo. In case you have some „dirt" on the ground because of a non-optimal layout, this little add on can avoid that ground contamination enters the differential output stage.

Talking about the differential nature of the INA we must talk about caveats too. Firstly, for best common mode rejection it is necessary that all resistors are well matched. 0.1% is fine I think for audio. In critical medical and aerospace applications 0.01% is not uncommon and resistors need to have a low and similar Tc.

The same matching requirement is true for the differential RIAA I use as well so frequency selective components like caps have to be matched too for best results. I will talk about that later when we discuss the differential RIAA equalization stage. The INA by nature has a balanced input so that means that is has 6dB more gain than a unbalanced solution but also has more noise due to the two Opamps in series plus the Johnson noise of R2. In total that amounts to a 3dB noise penalty compared to a single ended solution with the same amount of active input devices and the feedback resistor. For a more in depth analysis of noise in INAs and other topologies read Burkhard Vogel's book „The Sound of Silence". But fear not, in the proceedings of this article I will show a way to minimize noise close to the theoretical minimum. In fact the circuit simulates in Microcap to have -76dB of noise unweighted referenced to 1mV and that translates into better then 80dB weighted. No vinyl source material I know of has that kind of dynamic range. (Burkhard Vogel's book also discusses dynamic range) . He makes a convincing case that maximum dynamic range on a Vinyl record cut with DMM is around 72dB. A conventional lacquer cut vinyl record has a noise floor that is 6 to 8dB worse.

In a practical situation the Johnson noise from the impedance of the cartridge and the wires say 5 Ohm has to be added and even a theoretical noiseless MC input can not be better then 0.2nV/rtHz with a 5 Ohm cartridge connected. So in praxis we have the following situation: The lowest noise phonostages I know of are the Synaestesia HP3, HP4, HP5 and my Ultra Low Noise Rosi plus the Ultra Low Noise High Z MPP. I do not know any commercial stage on the market that betters this. Jürgen Ultee's Grail is in the same ballpark, so is John Curls Orion with 0.4nV/rtHz. Actually going lower then 0.3nV/rtHz is nearly impossible without substantial cooling due to the laws of thermo dynamics no matter how much devices you parallel, not to mention layout and oscillation problems. So the HP3 would be 0.3nV/rtHz plus 0.2nV/rtHz from the cartridge amounting to 0.5nV/rtHz

and the Discrete INA FPS I will present here is 0.5nV/rtHz plus 0.2nV/rtHz amounting to 0.7nV/rtHz, shrinking the theoretical 4.5dB advantage down to less than 2dB in a practical setting. I can live with that considering that the FPS has the advantage of balanced operation, namely superb rejection of external noise fields like hum from transformers and motors. So the FPS holds the low noise record for balanced active stages as far as I know.

But let's talk further how the circuit developed. I had build the "Hybrid" INAs out of separate Opamps. Still I was not totally satisfied with the noise performance. In the best case of the Low Z version two Opamps are in series with the signal so with the best devices on the market like the OPA1611, OPA211, AD797, LT1115, LT1028, LME49990, ADA4898 (yes, I tried them all) it did not get any better than 1.45nV/rtHz. The next step was paralleling and in this case I was successful. No oscillation like the one that my friend had encountered with the INA163. The most extreme case was coined „Dolphin" and had 4 plus 4 AD797 in parallel. It had a performance of 0.7nV/rtHz, very good for balanced, very expensive to make (those low noise chips are not cheap) and still not the theoretical minimum. Guess what, the next step was going discrete in the input stage so my Discrete INA was born. I decided on a cascoded FET input stage. The devices I chose are optimized for MC duty and are still available NOS although not produced any more. They are the Toshiba 2SK170BL and 2SC74BL. Linear Systems produce an N channel substitute and a P channel is on the agenda, so let's hope that these excellent designs are not lost forever. The same is true for low noise BJTs. Most of them are not produced any more but some medium power devices that are not explicitly designed for MC inputs are still available new (see www.synaesthesia.ca).

FETs have quite a high and nonlinear input capacitance and cascoding them with BJT's reduces input capacitance much. It also has the advantage of lower distortion because the FETs are driven at nearly constant voltage and current. PSU suppression is also better with cascoding and I feed the cascode from a well designed constant current source that reduces PSU ripple even more. The constant current sources can be adjusted so voltage at the collectors of the cascode BJTs can be held constant and equal in the positive and negative amplification chain. Voltage at that point should be set at around half of the PSU voltage to optimize dynamic range. For a + - 12V supply, 5 to 6V is fine. The input FETs are held under 3V then and this is not ideal for lowest distortion (7V would be better) but for lowest noise this is the best solution because under 3V no hot carriers develop that give rise to flicker noise. The distortion is taken care of in the feedback arrangement.

The cascode and current source BJT's are the old faithful 2N4401 and 2N4403. They are not bad in this application at all because they have low base spreading resistance (Rbb = 40 Ohm, extremely low values are not necessary here because noise in the cascode is not correlated with the input FETs) and low output capacitance. Best of all they are still made today in quantity and very affordable. They do not have high Hfe but in a cascode this has the advantage of high early voltage. When you are into swapping mode and want to try alternatives, see the sidebar. Hfe and Vbe matching is not

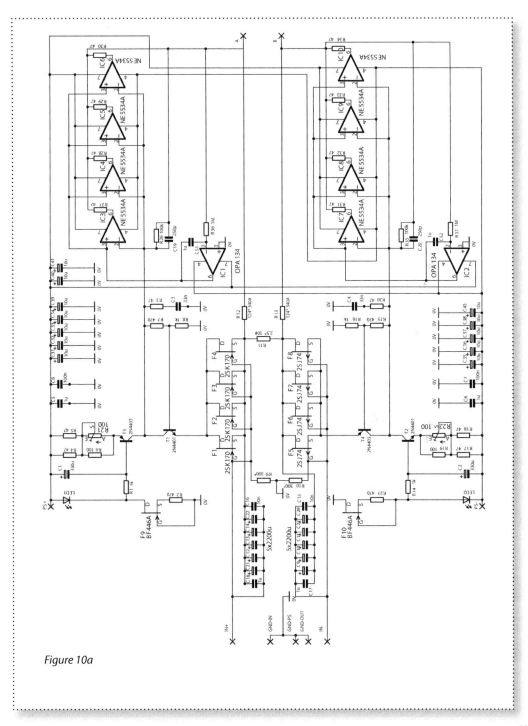

Figure 10a

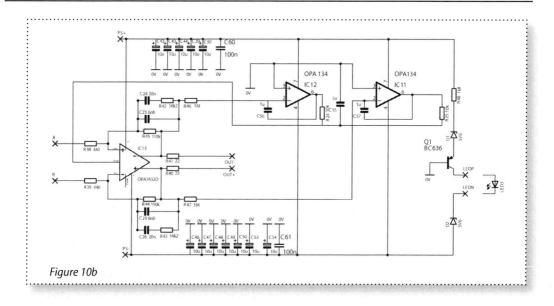

Figure 10b

tages: simplicity, low noise because a separate bias source with potential noise is not necessary and a measure of current feedback that stabilizes the gain, lowers distortion and raises the bandwidth a bit at the cost of less gain. Actually I have not tried the alternative separate bias version because the circuits worked beautifully right away, so if you think seperate bias is better, try it at your own risk. Be careful with the lead lag compensation though. Without the current feedback action of the ground referenced bias the already tremendous gain goes up sky high and other components may be necessary for compensation to avoid wild oscillation. Considering how long I needed to compensate the circuit for stability I recommend this praxis only for experienced builders that have some theoretical background in feedback theory. I found that this circuit was somehow hard to simulate so the chosen values where optimized by feeding a scope with a differential signal from the two shunt feedback Opamps. The values given result in a perfect 10 kHz square wave without any overshot. Frequency response in this arrangement is limited to 200 kHz -3dB with a shallow slope to avoid HF interference.

The next stage after the input plus lead lag compensation consists of Opamps in shunt feedback arrangement. They perform a current in - voltage out transformation somewhat limited by the resistors in the ground referenced common base stage. You may ask why I use 4 Opamps in parallel. The reason is that we go for the lowest noise and that means we have to make R2 small so when we want a gain of say 36dB R1 gets small too. Besides making noise low I also like to make the overload margin as high as possible. An average Opamp can drive 20mA clean so 4 of them can drive 80mA. Brute force paralleling is usually not recommended but putting 47 Ohm resistors at the output isolates the Opamps well enough and this circuit works without problem. Each leg can drive 9.92V in a 124 Ohm

resistor so we get 19.84V total after the differential stage. This is an insanely high overload margin. If we use a cartridge with 0.5mV output at 5msec and we assume that 50m/sec can be cut at certain frequencies we have a maximum input voltage of 5mV. Add to this a gain of 40dB for example and the maximum output voltage is 500mV. That is a relation of 19.84 / 0.5 = 39.68 leading to an overload margin of 32dB. 50m/sec does not exist in reality but some records may have surface blemishes like scratches and dirt that can send transient rich high energy pulses into the system. John Curl measured transients up to 500 kHz. Anyway, I found that phonostages with high overload margin sound better and can handle scratches much better making them audibly „shorter" and separating them in the sound field from the music. The scratches coming from the left and right loudspeaker plane and the music stand free in 3 dimensional space provided your speakers are up to the task. You may ask why I do not use one Opamp with higher current drive ability like the OPA551. The reason is that my arrangement is more flexible in Opamp choices. So if you do not like to use the NE5534A you can put in your favorite specimen. I tried LME49710 and OP27 with success so the more modern OP227 should work too. If you have a weak spot for FET designs you can try OPA134, OPA1641 or the even better spec'd OP827 and ADA4627. Be careful when the Gain-Bandwidth Product goes much over 20MHz because the circuit could start to oscillate. OPA827 can be used in the Servos too. DC offset will then go down from 1mV of the OPA134 to less than 0.1mV.

Coming back to the function of the circuit the result is tremendous gain that Allen Wright of Vacuum State called "scary". Again I am using an old and well know part here: the NE5534A, this a selected version (Noise + DC offset) of the NE5534 that started life more than 30 years ago as the TDA1034 from Signetics. Why this old war horse? First it is a very low distortion part, even today and I use only the basic transfer linearity that is excellent in the shunt feedback arrangement used. Second it has low current noise for a BJT input Opamp, 4 times less then the more recent LME49710. This is necessary for low noise in the current to voltage transformation. A FET Opamp could be used as mentioned but results were fine. Third it has a 10MHz gain – bandwidth product and this just right for the compensation used. The faster LME49710 worked too but I did not have success with the very fast AD797 for example. It (the NE5534) has some issues with DC offset but the servos take care of that and the 5 to 6 volt bias from the input stage. A compensation cap is fitted over the 100kOhm resistor for stability and optimum time domain behavior. Feedback resistors both from the positive and the negative side go to R2 and our discrete INA is ready. Well simply combining the outputs would not work because we would only hear the common mode signal that is superbly low. I know it because I tried it. What we need is a differential stage that extracts the amplified input signal. It occurred to me that this stage could also do the RIAA differentially. That would save an additional output stage and a differential RIAA done right has a lot of advantages too.

This time I use a very modern part, the OPA1632. It is differential in – differential out and ideal for this purpose. It has very low voltage noise, low distortion and high bandwidth and can drive low impedance. A differential stage could be build with separate Opamps too but I prefer this integrated

solution. Layout can be done very compact and performance is excellent. For the RIAA I use a shunt feedback arrangement. It has many advantages that where described by John Linsley Hood in his book „Valve and Transistor Audio Amplifiers". I simply will quote him here and add some commends: „Shunt feedback layouts can also give a precise match to the RIAA specification, which a series feedback circuit, as it stands, will not. This is because any series feedback layouts having a feedback limb which decreases in impedance with frequency will tend to unity gain, whereas the RIAA specification stipulates a gain curve which tends to zero gain. Unfortunately, in many preamplifiers of commercial origin, which use series feedback RIAA stages in order to obtain lowest practical (input short-circuited) noise levels, this error is ignored, even though its effect can easily be seen on an oscilloscope and the tonal difference it introduces can easily be heard by the listener." Noise from the shunt feedback resistor in my design is low because of the head amp architecture with a lot of gain before the RIAA. As JLH continues, "The effect of this is to alter the relative character of the noise given by these two systems, so that the noise from a series-type circuit is a high pitched hiss, while that from a shunt-type circuit is more of a low pitched rustle." And finally: "There are also a few other advantages offered by shunt-feedback RIAA circuits. Of these, the most important, in practice, is that they are largely immune to the very short duration input overload effects, which can be caused by voltage spikes in the cartridge output arising from dust and scratches on the surface of the record. These short duration overloads can greatly exaggerate the clicks and plops which are a nuisance on vinyl disc replay. Also, they have, in theory, a lower level of distortion"

Here we have the overload theme again. Some beta testers of my design suggested a passive RIAA at least for the treble 75usec time constant. RIAA precision with passive is as good as shunt-active but I refused because that comes with an overload penalty. Even doing only the treble passive would worsen the overload margin by 14dB and increase distortion due to less feedback in the treble. See D. Self's book "Small Signal Audio Design" for an in depth discussion about the advantages of active RIAA stages. What is true for transistor design may not be true for tube design. Tubes running on very high voltage have a huge overload margin by default so I would see no problem designing a good sounding passive RIAA with tubes but that is not my forte and a bit off topic. The differential nature of the RIAA has also the advantage that component tolerances average out somewhat but for best common mode rejection caps and resistors should be matched. 0.1% resistors are recommended and easy to get, though a bit more expensive. Good quality caps with 1% tolerance are available in Polypropylene and Styroflex and actually the values chosen give 0.1dB RIAA precision with 1% parts. For best common mode rejection this is not good enough. 1% tolerance gives a CMR of 40dB, a bit meager but still useful. I would suggest that you select quadruples of caps for the positive and negative chain and the right and left channel. It is not so important that they have the absolute correct value, 1% off gives still + - 0.1dB accuracy, but that both channels and plus and minus chain are similar. CMR will benefit, so does 3 dimensional imaging and focus. The FPS finishes off with two 22 Ohm resistors that drive the balanced output and a differential servo that holds the DC offset between non-inverting and inverting output low.

So here we have it: A thoroughly modern State of the Art Phonostage that you can build with parts that are available. I can not hear a loss of information with this circuit, instead the sound is very natural, clean and dynamic but your ears may tell you something different. At least from the standpoint of objective performance this stage is hard to better.

Closing notes

You need a balanced to unbalanced converter to connect this stage to an unbalanced input. A passive solution is possible too and Jan Didden has some tips for that at www.linearaudio.nl . A suitable power supply would be one of the variants of the Didden/Jung super-regulators, published by Audio Amateur in 1995. Reprints are available at Jan's site www.linearaudio.nl and from Walt Jung's site www.waltjung.org . Finally, a PCB and parts kit will be available from Pilgham Audio (see layout at the end) at www.pilghamaudio.com , who can also provide power supplies.

Thanks to Allen Wright, Salas, Syn08, Ward Maas, Sigurd Ruschkowski, Jan Didden, Scott Wurcer, John Curl and others that contributed in one or the other form. I learned a lot from the books written by Douglas Self, Burkhard Vogel and John Linsley Hood. Without all that help, this project would not exist. For the others not mentioned, you know who you are.

Component swapping

1 : 2N4401 and 2N4403 can be substituted by BC550 / BC560,
2SA970 / 2SC2240 or 2SA1015 / 2SC1815 (but watch different pin layout);
2 : NE5534A can be substituted with LME49710, OP27, OPA227, OPA134, OPA1641;
3: Resistors: 0.1% or better, metal film, low noise, low tempco;
4: Electrolytics: Panasonic FM or similar for bypass, Nichicon LG for coupling;
5: OPA134 in the servos can be substituted with OPA827 for lowest offset (0.1mV);
6. Bypass Caps: Wima MKT, Epcos MKT, Rifa SRM or similar;
7. RIAA caps 1% tolerance, matched in positive and negative chain, ideally to 0.1%.
Recommended types: LCR Styroflex, Roederstein MKP1837, 1% Silver Mica.

Configuration options

The name FPS comes from the fact that you can arrange the input in two different ways: High Z and Low Z. The High Z version has the advantage of DC coupling and loading resistor types and values can be changed at will to optimize coupling to the cartridge. Low values usually sound more robust and high values sound more airy. I use 2 x 300 Ohm on my Lyra Titan I for example. Higher value loading gave a sound that I found too bright but your system and taste may differ.

The Low Z version needs AC coupling. I use very modern shunted electrolytic caps with polymer dielectric here. The Nichicon LG series I use has a 2200uF cap that can stand more than 6A of ripple current and has an ESR @ 100 kHz of only 8mOhm. It comes in a very small package so a compact layout is possible. It's disadvantage is a high leakage current but in my design this current cancels in the cartridge because the current has opposite phase. Actually I could not measure any DC at the input with the meters I had available. Bias voltage is small over them and I use 5 plus 5 in parallel so distortion down to low frequencies should be small. Contrary to popular believe, bias raises distortion in electrolytics. A general rule is that less voltage over the electrolytic gives less distortion. Another advantage of the Low Z version is that the sources of the FETs are driven, so the non-linear input capacitance of the FETs that is not in the feedback loop is out of the equation. Theoretically the Low Z version is faster and has less distortion in the treble. In terms of noise the Low Z version has the edge. Impedance of the cartridge is parallel to the feedback resistor R2 in this case so that resistor actually lowers the noise of the cartridge. For example a feedback resistor of 5 Ohm plus the 5 Ohm of the cartridge would make the noise impedance 2.5 Ohm. Compare that with the 7.5 Ohm of the High Z version. If we would raise the feedback resistor to 10 Ohm the noise impedance would still be lower then the High Z case and the resistors would provide a measure of source degeneration that would improve linearity of the input stage. An endless game of permutations. Sound wise I heard the same differences I had observed with my bipolar designs. The High Z version has very elegant tonality with some measure of warmth and the Low Z version has tremendous bass precision with a lot of spectacular raw dynamics. It's up to you.

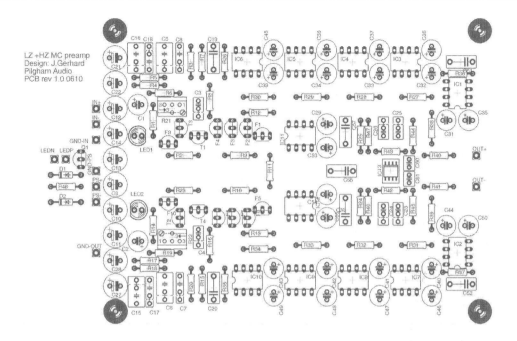

LZ +HZ MC preamp
Design: J.Gerhard
Pilgham Audio
PCB rev 1.0 0610

A New Low-Noise Circuit Approach for Pentodes

Frank Blöhbaum

This article is based on a presentation which was given at the European Triode Festival (ETF) 2009. The investigation covers tubes and circuitry for preamplification, where low noise behaviour is a must.

Why Pentodes?

At first sight it seems stupid to spend time for investigating pentodes for tube based preamplifiers having as lowest noise as possible. If one takes a look into the web one will find that almost everybody knows for shure that:

a)   Pentodes are much more noisy than triodes

b)   Pentodes have much more odd harmonics than triodes

That is not just a myth. Even very well respected tube experts like Morgan Jones have warned to use pentodes for low noise amplification. He writes {1} that pentodes have typically between 6dB to 14dB more noise than that pentode connected as a triode!

BUT, if one tries to get hard facts like noise values in the datasheet of a given tube or real measurements you will very likely find – nearly nothing. There are some noise values for high frequency (MHz) range, but these are almost useless for low frequency amplification because of the 1/f-noise which is dominating in the audible spectrum of the noise. As an example in {2} the equivalent noise resistance of Philips E180F is given with typically 460 Ω, max. 650 Ω – for HF use. The given equivalent noise resistance of the E810F is extraordinary low: 110 Ω, but for HF use only. One rule of thumb is obvious: the higher the transconductance the lower the noise of the tube. That is true for both triodes and pentodes. Following this insight there should be alternative options for the common low gm audio tubes like ECC83 and EF86: high-gm low noise pentodes.

First lets have a look how much gain is theoretically achievable with triodes and pentodes:

Pentodes: $G = gm * Ra$ [1]

Triodes: $G = \mu * Ra / (Ra + ri)$ [2]

If one uses a constant current source for Ra with say 100 kΩ impedance the gain of a typical high-gm pentode like E180F is:

G (E180F) = 16,5mA/V * 100 kΩ = 1650;

and for one ECC83 having ri = 62.5 kΩ and μ = 100:

G (ECC83) = 100 * 100 kΩ / (100 kΩ + 62,5 kΩ) = 61,5.

The gain of triodes is always limited by the voltage amplification factor μ, which is in the range of 10 ... 100. So pentodes offer a compelling feature: in theory the voltage gain of one grounded catode amplifier stage is unlimited for unlimited Ra. In practice high-gm pentodes combined with moderate anode impedance should excel triodes in terms of gain – and maybe in terms of noise as well – which has to be evaluated.

How should we compare noise performances?

Lets have a look into the datasheet of two famous examples for low noise audio amplification in the past. By the way these are one of the few examples where noise data for audio use has been provided. The Philips datasheet states, for an EF86 in Pentode mode:

„The equivalent noise voltage on g1 is approximately 2μV for the frequency range from 25 to 10000 Hz at Vb = 250V and Ra = 100 kΩ"

For comparision lets take the ECC808, which is a special high quality low noise version of ECC83 / 12AX7 triode. The Telefunken datasheet tells us:

„At Uba = 250V, Ra = 220 kΩ the equivalent noise voltage across the grid is approximately 2μV for the frequency range 45Hz to 15kHz."

So at first look both tubes might have the same noise voltage at the input => 2μV. BUT, the given bandwidth for this noise voltage is significantly lower for the EF86: 10000-25 = 9975 Hz instead of 15000-45 = 14955 Hz for the ECC808. Therefore the ECC808 has the lower noise than the EF86 at the same measured bandwidth.

For comparing noise we have to use a bandwidth normalized noise value:

Equivalent Input Noise Density en [V / RtHz]

Lets do the math for the two datasheet values of EF86 and ECC808:

en (EF86_datasheet) = 2μV / (9975)$^{1/2}$ = 20,025 nV/RtHz   [3]
en (ECC808_datasheet) = 2μV / (14955)$^{1/2}$ = 16,35 nV/RtHz   [4]

But how could we get these numbers in practice by measurement? Measuring the input noise voltage is extremely difficult. So we would measure the output noise voltage of an amplifier at a given and constant measurement bandwidth. As an example we would have measured two amplifiers: An amplifier #1 having a gain of 100 and an output noise voltage of 100µV and a second amplifier #2 having a gain of 20 and an output noise voltage of 50µV. The equivalent input noise voltage of amplifier #1 will be 100µV / 100 = 1µV and that of amplifier #2 will be 50µV/20 = 2,5 µV.
It`s now obvious, that the amplifier #1 is much less noisier than amplifier #2.

So for comparing noise we have to take into account the bandwidth as well as the gain of the amplifier under test. The best way to do this is to calculate the Equivalent Input Noise Density.

Partition Noise

Regarding noise sources the main difference between triodes and pentodes is partition noise. Fig. 1 shows the effect of increased noise by partition noise:

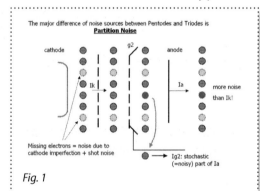

Fig. 1

The positive connected screen grid of the pentode attracts a part of the cathode current I_k. This screen grid current has a randomized number of electrons in a given time slot. That causes randomized „missing places" of electrons in the former cathode current stream – which will be now the anode current. Consequently the anode is hit by more stochastic fluctuating electrons = more noise.

What is happening inside a triode connected pentode? Fig. 2 shows a rough explanation, why the noise of a triode connected pentode should be lower than that of the same tube connected as a pentode:

The cathode current is still partitioned into the screen grid current and anode current. But at the anode connection the screen grid current is reunited with the anode current. If the wire from the screen grid connection to the anode connection is short there should be no significant delay in the audio frequency range. Therefore the electrons could take their place again inside the anode current stream and the

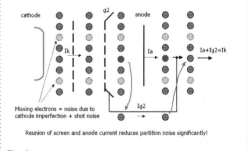

Fig. 2

partition noise is eliminated. Of course, this is a very simplified model and experts in quantum physics will tell you that this is definitely not the complete story... At least the additional shot noise term of the screen grid current should have an effect. But it is an easy understandable explanation for a complicated theory.

So it is obvious that the maximum improvement of noise values for a given pentode could not be higher than the ratio of I_{g2} / I_a. The typical ratio of I_{g2} / I_a is between 1/5 and 1/3. That`s the reason why I doubt very much the improvement of „6dB to 14dB" if a pentode is connected as a triode.

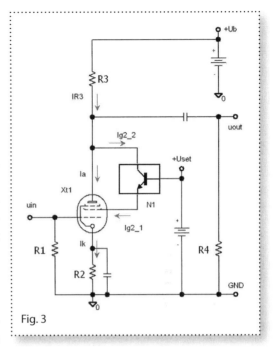

Fig. 3

New BestPentode Circuit

If we connect a pentode into triode mode we might reduce the noise a bit, but we reduce the gain a lot as well. In triode mode the μ_{g2g1} is now the absolute limit of achievable gain.

On the other hand the classic pentode circuitry has the additional drawback that the screen grid current is not used for amplification – a significant part of the cathode current is wasted.

A better circuitry for pentodes should have these features:

- *applicable to almost any pentode but favourably to high-gm types $G = gm * Ra$*
- *Screen grid fed by fixed voltage supply having low impedance pentode mode*
- *Reunion of screen grid current I_{g2} and anode current I_a no partition noise*
- *Use of otherwise wasted I_{g2} for amplification for more gain than classic pentode mode*
- *Independent of I_{g2}/I_a split ratio for better linearity*
- *Should be easy to implement*

In Fig. 3 the new pentode circuit is presented having the features listed above:
In this figure the potential +Uset is buffered by npn transistor N1 which, via its emitter, sets the screen grid voltage of pentode Xt1. The base current is so low that it is negligible. If we take the physical right direction of the current flow – from „-" to „+" - the screen grid current flows into the emitter of N1 (I_{g2_1}, appears back at the collector of N1 (I_{g2_2}) and is then reunited with the anode cur-

rent: $I_a + I_{g2\_2} = IR_3$. This current IR_3 flows through R3 which represents one part of the anode working load R_a. The other one (for ac-signal) is the input impedance of the following amplifier R4. So for the gain calculation formula we have to consider that Ra equals the parallel connection of R3 and R4:

$$Ra = R3 // R4 = R3 * R4 / (R3 + R4) \qquad [5]$$

Both currents, I_{g2} and I_a, are working now for signal amplification, I_{g2} is not wasted anymore. Furthermore, the partition noise should be reduced significantly because I_{g2} and I_a are reunited again and the „bypass" current path of N1 has a large bandwidth. Because the screen grid voltage is fixed the gain of Xt1 equals still to equation [1] – but gm is higher because of the additional working of I_{g2}.

N1 could be realised with different active parts: npn bipolar transistor, n channel MOSFET or an additional tube. The bipolar transistor might be the best solution because it`s cheap (for example <10 cents for MPSA44) is very easy to implement and has the lowest noise.

Because the main features of the pentode are improved – gain and noise – and the pentode is still working like a pentode I called it the „BestPentode" circuit. Please take note that this circuit is appropriate for small signal amplification and driver stages. The anode voltage must always be higher than the screen grid voltage. It is not useful for power amplifier stages where the anode voltage is much lower than the screen grid voltage at full output power.

Measurements
1) Measurement setup
To validate the theory I set up a special low noise measurement procedure.
It is very important to remember that pentode circuits have a power supply rejection ratio of zero dB – any noise and/or hum of the power supply will be part of the output signal! If we take the EF86 data for example we would expect an input noise value of about 2...3µV multiplied by a gain of approx. 100 which delivers 200...300 µV output noise voltage. In triode connection the output noise voltage is even smaller because of the reduced gain. We could expect something down to 20...30µV. So the power supply must have a noise residual not higher than 2...3µV.

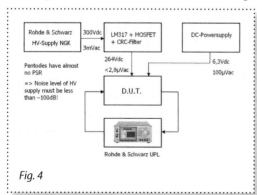

Fig. 4

Fig. 4 shows the measurement setup:

The Rohde & Schwarz high-voltage supply device NGK is used for generating +300V. This voltage is postregulated with a floating LM317 which is buffered by a MOSFET. After that a CRC filter cascade is supressing the hum and noise

further down to less than 2,8µVrms at a bandwidth of 20Hz – 22kHz.

The post regulator and CRC-Filter is located insight a shielded box which includes the amplifier circuitry as well.
The noise and gain is analyzed with the UPL precision audio analyzer from Rohde & Schwarz.
I defined some principles for the measurements:

- Goal: Verification of the advantages / disadvantages of circuit structures, not tube types!
- No optimization of individual working points of the tested tubes for best individual noise or linearity properties (no tweek for best individual results): Use of one and the same parts for different circuits to have one and the same conditions for all tests. Exception: measurements of EF86 and ECC808 because of their much different requirements.
- Different tubes having different construction have been used (i.e. normal grid; frame grid; low, medium and high transconductance gm), but not enough tubes from each type were tested to have a statistical approved answer which tube is best or worse. No buying recommendation !!
- Noise measurement is done with a 50Ω plug at the input BNC-connector. The RMS noise value is calculated via FFT-spectrum and Post-FFT calculation done by the UPL.
- THD-measurements were done with normalized output voltage of 1 Veff = 1,414 Vpeak and a Kaiser window. The input stimulation is done by the UPL generator.
- Bandwidth is 20Hz – 21,938kHz for all measurements (given by R&S UPL).
- Because usage of high-gm frame-grid pentodes is most interesting, some high-gm frame-grid triodes are tested for comparison as well.

2) First validation of classic EF86 and ECC808 circuitry

For validating the measurement setup the tubes which have data sheet values for noise were used first: EF86 and ECC808. The working points follow the recommendation of the data books. Fig. 5 shows the recommendation for EF86 in pentode mode.

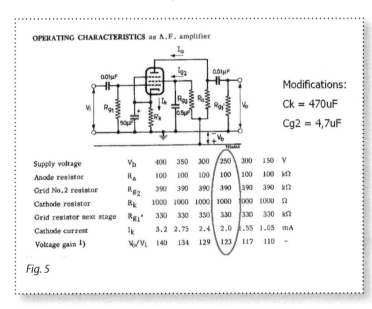

OPERATING CHARACTERISTICS as A.F. amplifier

Modifications:

Ck = 470uF

Cg2 = 4,7uF

Supply voltage	V_b	400	350	300	250	200	150	V
Anode resistor	R_a	100	100	100	100	100	100	kΩ
Grid No.2 resistor	R_{g2}	390	390	390	390	390	390	kΩ
Cathode resistor	R_k	1000	1000	1000	1000	1000	1000	Ω
Grid resistor next stage	R_{g1}'	330	330	330	330	330	330	kΩ
Cathode current	I_k	3.2	2.75	2.4	2.0	1.55	1.05	mA
Voltage gain [1]	V_o/V_i	140	134	129	123	117	110	–

Fig. 5

51

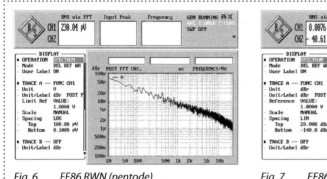

Fig. 6   EF86 RWN (pentode)

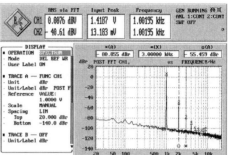

Fig. 7   EF86 RWN (pentode)

The measurement of the output noise spectrum of EF86 RWN (made by Röhrenwerk Neuhaus RWN in East Germany) in classic pentode mode is shown in Fig. 6.

The measured output noise voltage is 230,04µVeff. Please take note of the remarkable high 1/f-noise! The corresponding THD-spectrum, which is used for gain calculation as well, is shown in Fig. 7. The 0dBr-point equals to 1Veff = 1,414Vpeak. The measured input and output voltage is shown in the middle section „Input Peak". The gain for the EF86 (RWN) is:

$G = 1,4187V / 13,183mV = 107,62$

Now we can do the math:

$en (EF86RWN\_pen\_measured) = (230,04µV / 107,62) / (21918)^{1/2} = 14,438\ nV/RtHz$ [6]

Compared with the databook value [3] that is a better than expected value. But other measured EF86 were much worse: three NOS EF86 made by Telefunken have 31,7nV/RtHz ; 44,01 nV/RtHz and a wopping 60,24 nV/RtHz respectively!

Lets go to the triode connected EF86. The circuit is drawn in Fig. 8, the suppressor grid g3 is not

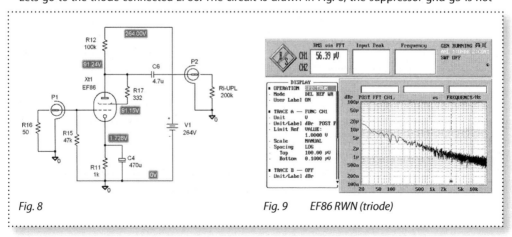

Fig. 8

Fig. 9   EF86 RWN (triode)

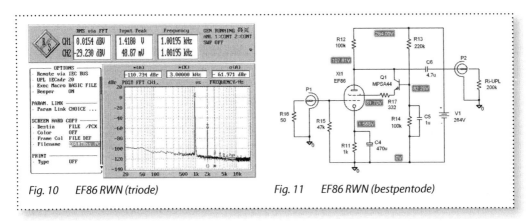

Fig. 10   EF86 RWN (triode)   Fig. 11   EF86 RWN (bestpentode)

shown and connected to screen grid g2. R17 is used for suppression of VHF oscillations. Unfortunately neither the datasheets of Philips nor those of Telefunken tell us anything about the noise values for a triode connected EF86. So lets have a look to the real measurements. Fig. 9 shows again the output noise spectrum of EF86 (RWN), now in triode connection:

The 1/f-noise characteristic is the same. The measured output noise voltage is now 56,39µVeff. Fig. 10 shows the matching THD spectrum for 1Veff output signal:

In triode connection we need now an input signal of 48,87mVpeak to get an output voltage of 1,4180Vpeak. The gain is therefor 29,02 only. Please take note the much lower k3 compared with the pentode connection!
The noise calculation delivers:
en (EF86RWN_tri_measured) = $(56{,}39\mu V / 29{,}02) / (21918)^{1/2}$ = $13{,}125\ nV/RtHz$ [7]

Comparing the measured noise of the triode connection [7] with that of the pentode connection [6] we see an improvement of about 9,08% or 0,83dB only for the triode mode. That is an improvement but it is much less than the 6...14dB expected in {1}.

The next step goes into unknown territory: the measurement of the BestPentode mode. The schematic is shown in Fig. 11, the suppressor grid g3 is again not shown and connected to g2:

The screen grid voltage is set via emitter Q1 to 81,7 Volts. That is lower than the recommendation for the classic pentode mode datasheet, but required to have some headroom to the anode voltage. The measured noise of the BestPentode mode for EF86 RWN is shown in Fig. 12:

The 1/f-noise characteristic remains the same. The measured output noise voltage is now 194,15µVeff. Fig. 13 shows the matching THD spectrum for 1Veff output signal:

In the new BestPentode connection we need now an input signal of 11,377mVpeak only to get an output voltage of 1,4203Vpeak. The gain is therefor:
$G = 1,4203V / 11,377mV = 124,84$

k3 is significantly lower than in classic pentode connection. That could be expected because the nonlinear partition ratio between I_a and I_{g2} is irrelevant in the BestPentode mode.
The noise calculation delivers:

en (EF86RWN_BP_measured) = (194,15μV / 124,84) / (21918)$^{1/2}$ = 10,504 nV/RtHz [8]
Comparing the measured noise of the BestPentode connection [8] with that of the classic pentode connection [6] and the triode mode [7] we see that the BestPentode mode for EF86 is by far the best

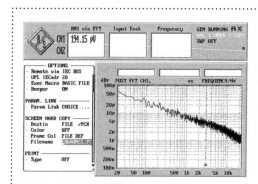

Fig. 12   EF86 RWN (bestpentode)

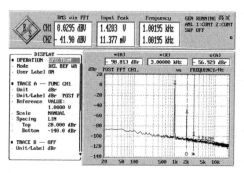

Fig. 13   EF86 RWN (bestpentode)

one in terms of noise. The much better noise figure in comparision to the triode mode is surprising. Additionally the gain is the highest and the k3 is 18dB lower than that of the classic pentode mode.

Finally we will have a look at the noise of the special low noise triode ECC808. The part values follow the recommendation of {4}: Ra = 220kΩ, Rk = 1,7kΩ. The measurement was done with 2 sections of ECC808 connected in parallel, because of that the cathode resistor Rk was lowered to 845Ω and the anode resistor to 110kΩ. The measured noise spectrum of NOS ECC808 made by Telefunken is shown in Fig. 14.

The 1/f-noise is lower than that of the EF86. The measured output noise voltage is 45,75μVeff – for 2 triodes in parallel.

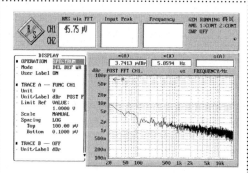

Fig. 14   ECC808 NOS TFK

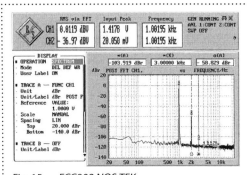

Fig. 15   ECC808 NOS TFK

Fig. 15 shows the matching THD spectrum for 1Veff output signal.

The gain of the 2 parallel connected ECC808 triodes equals to:

$G = 1,4178V / 20,05mV = 70,71$

For the noise calculation we have to consider the 2 parallel connected triodes. So we have to multiply the measured noise value by a factor of SQRT(2) = $2^{1/2}$ = 1,4142 to get the value for one ECC808 triode:

en (ECC808_measured) = (1,4142 * 45,75µV / 70,71) / $(21918)^{1/2}$ = 6,180 nV/RtHz   [9]

This noise value is much better than the one in the databook [4] and excels all circuits of measured EF86.

3) Validation of medium- and high-gm tubes

Medium and high transconductance (gm) pentodes were measured in 4 different test circuits each, presented in Fig. 16:

The testcircuit Fig.16b) was chosen for having identical working points for the pentodes independent of their actual screen grid current. Additionally the comparision of 16a) and 16b) could give an indication about additional noise of the added npn-transistor and the comparison of 16b) and 16d) will give an indication about the advantages of the BestPentode circuit against the pure pentode circuitry.

These tube types were measured:
- 8 x different high-gm frame grid pentodes: CV6189 (GB), D3a (Siemens), E180F (Valvo), E186F (Valvo), EF184 (Valvo), 6Э6П-E (RUS), 6Ж9П (RUS), 6Ж11П-E (RUS)
- 2 x medium-gm pentodes: EF80 (Valvo), E81L (Valvo)
- 2 x high-gm triodes: EC806S (Siemens), EC806S (Tfk)

The measurement results are presented in the following tables below. Of course the already mentioned EF86 (RWN), EF86 (Tfk) and ECC808 (Tfk) are included in these tables for better comparison as well.

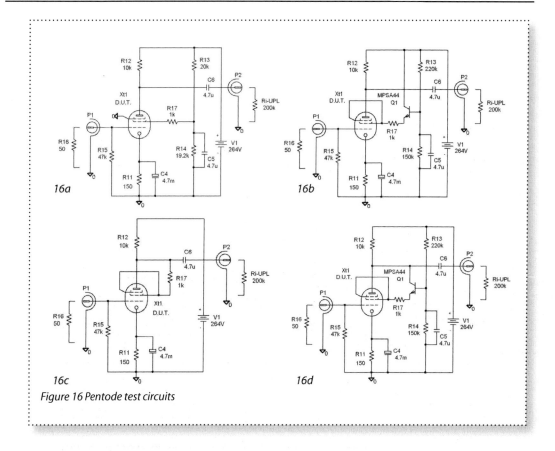

Figure 16 Pentode test circuits

Table 1 shows the results for the equivalent input noise density.

Some circuit designers like more the noise representation in terms of equivalent noise resistance which is shown in Table 2.

The measured gain is shown in Table 3.

The results for the measured first two harmonics k2 and k3 at a standard output voltage of 1Vrms = 1,414Vp is shown in Table 4.

	Classic Pentode	EF-Pentode	Triode	BestPentode	Winner Noise
6Э6П-E	3,64	3,76	3,69	3,43	BestPentode
CV6189 GB	6,46	6,5	7,81	6	BestPentode
D3a Siemens	2,66	2,65	2,44	2,4	BestPentode
E186F Valvo	3,81	3,68	3,8	3,4	BestPentode
6Ж11П-E	4,95	4,93	5,03	4,67	BestPentode
6Ж9П	3,31	3,42	3,06	2,92	BestPentode
EF80 Valvo	7,77	8,34	10,27	7,74	BestPentode
EF184 Valvo	4,92	4,43	3,96	3,45	BestPentode
E81L Valvo	4,59	5,23	4,2	3,85	BestPentode
E180F Valvo	5,62	5,56	5,72	5,32	BestPentode
EF86 RWN	14,44	11,93	13,125	10,504	BestPentode
EF86 Tfk	31,7	19,71	25,78	17,82	BestPentode
EC806S Siemens			9,86		
EC806S Tfk			3,28		
ECC808 Tfk			6,18		

Table 1: Measured Equivalent Input Noise Density [nV/RtHz]

	Classic Pentode	EF-Pentode	Triode	BestPentode	Winner Noise
636П-E	801	852	823	708	BestPentode
CV6189 GB	2521	2548	3683	2173	BestPentode
D3a Siemens	427	423	360	348	BestPentode
E186F Valvo	875	817	870	698	BestPentode
6Ж11П-E	1479	1465	1528	1317	BestPentode
6Ж9П	660	705	565	515	BestPentode
EF80 Valvo	3643	4194	6372	3615	BestPentode
EF184 Valvo	1464	1185	947	718	BestPentode
E81L Valvo	1270	1650	1066	895	BestPentode
E180F Valvo	1907	1868	1977	1705	BestPentode
EF86 RWN	12585	8590	10398	6660	BestPentode
EF86 Tfk	60653	23448	40114	19167	BestPentode
EC806S Siemens			5846		
EC806S Tfk			648		
ECC808 Tfk			2305		

Table 2: Measured Equivalent Noise Resistance [Ω]

	Classic Pentode	EF-Pentode	Triode	BestPentode	Winner Gain
6Э6П-E	157,9	160,4	30,935	190,37	BestPentode
CV6189 GB	93,902	90,098	36,415	106,36	BestPentode
D3a Siemens	188	190,97	56,618	241,09	BestPentode
E186F Valvo	129,29	124,83	38,098	144,02	BestPentode
6Ж11П-E	122,42	118,85	34,927	133,84	BestPentode
6Ж9П	118,62	129,37	34,102	151,33	BestPentode
EF80 Valvo	64,61	67,056	33,213	82,934	BestPentode
EF184 Valvo	116,39	122,61	45,358	168,86	BestPentode
E81L Valvo	74,97	80,75	27,11	93,395	BestPentode
E180F Valvo	89,21	86,345	36,957	101,76	BestPentode
EF86 RWN	107,62	101,8	29,02	124,84	BestPentode
EF86 Tfk	95,78	90,53	29,31	108,15	BestPentode
EC806S Siemens			44,63		
EC806S Tfk			44,28		
ECC808 Tfk			70,71		

Table 3: Measured Gain [Uout / Uin]

	Classic Pentode		EF-Pentode		Triode		BestPentode		Winner k3
	k2	k3	k2	k3	k2	k3	k2	k3	
6Э6П-Е	-62,84	-91,47	-61	-92,8	-75,1	-105,9	-60,2	-97,7	Triode
CV6189 GB	-48,16	-108	-46,9	-111	-62,7	-114,7	-48	-108,5	Triode
D3a Siemens	-51	-95,4	-51	-94,5	-64,1	-93	-54,1	-99,3	BestPentode
E186F Valvo	-54,3	-86,3	-54	-83,2	-64,6	-86,8	-54,2	-83,7	Triode
6Ж11П-Е	-51,8	-95,2	-50,1	-110	-63,7	-107,5	-50,8	-109,2	EF-Pentode
6Ж9П	-51,7	-100,8	-52,5	-100,6	-66,8	-108,5	-53,6	-107	Triode
EF80 Valvo	-53,7	-92,5	-54,5	-88,8	-63,1	-84,6	-54,5	-87,7	Classic-Pentode
EF184 Valvo	-51,8	-85,9	-54,4	-92,6	-62,9	-103	-55,2	-108,7	BestPentode
E81L Valvo	-53,5	-102,6	-54,2	-96,5	-62,2	-101,5	-54,9	-100	Pen-R
E180F Valvo	-47,9	-93,9	-46,7	-97,5	-61,9	-97,6	-47,9	-100	BestPentode
EF86 RWN	-55,5	-80,9	-55,6	-92,1	-62	-110,7	-56,9	-98,8	Triode
EF86 Tfk	-61,5	-80,8	-55,9	-101,4	-63,3	-109,5	-58,2	-105	Triode
EC806S Siemens					-64	-109,8			
EC806S Tfk					-66,9	-108,1			
ECC808 Tfk					-58,8	-103,9			

Table 4: Measured Harmonics k2 and k3 [dB] referenced 0dB = 1Veff

The results of the measurements in Table 1 – 4 show, that theory and practice is not in all cases on the same track. So it`s surprising, that in all 12 cases the BestPentode circuitry has lower noise than the triode connection. The bottom line is that the partition noise of the classic pentode circuitry is completely removed by the BestPentode circuit. Please take note of the higher noise of the Emitter-Follower powered screen grid (EF-pentode) circuit compared with the BestPentode circuit. In both circuits one and the same transistor is used, but inside the BestPentode circuit I_{g2} is reunited again with the anode current, inside the EF-Pentode it`s not.

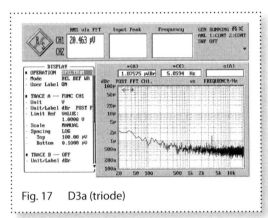

Fig. 17   D3a (triode)

Additionally the new BestPentode circuitry is the clear winner in terms of gain like expected. Pentodes having a large screen grid current in relation to the anode current like the EF184 have a lot more gain in BestPentode connection because the I_{g2} is not wasted anymore. In terms of harmonics structure the first picture is mixed. First surprise is that the pentode is not that bad like many prejudice are telling us. At a first look the characteristic of the spectra look the same – k2 is dominating, k3 is much lower and in most cases the higher harmonics are invisible. A more detailed look shows that the k2 battle is clear won by the triode, k3 is a mixed bag. The triode has the lowest k3 in 6 of 12 cases. But one should consider that the triode connection delivers much lower gain. If you spend a fraction of the much more gain of the BestPentode circuit for local feedback you will

very likely end up at the same level of k3 distortion or better. That is important for driver stages were a part of the output signal could be directly fed back.

Lets have a more detailed look at the noise spectrum of the first two winner pentodes in triode connection and BestPentode circuitry. Fig. 17 shows the one for the D3a in triode mode:

The 1/f-noise is quite low, but there is a small coloration of the noise at approx. 3,5 kHz.
Fig. 18 shows the noise spectrum of the D3a in BestPentode circuitry:

The 1/f-noise is very low in this connection as well. The coloration is more suppressed.

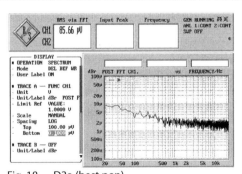

Fig. 18   D3a (best pen)

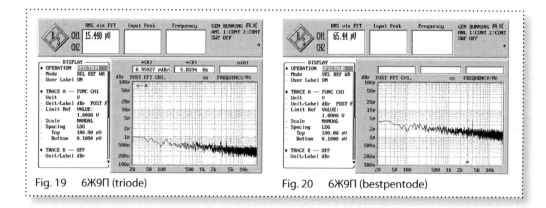

Fig. 19 6Ж9П (triode)   Fig. 20 6Ж9П (bestpentode)

The noise spectrum of the promising 6Ж9П pentode in triode connection is shown in Fig.19: The 1/f-noise is remarkably low. Please take note of the very low absolute value of the output noise of 15,44μV(!) only.

Fig.20 shows the noise spectrum of the 6Ж9П in BestPentode circuitry: The very good 1/f-noise characteristic is preserved.

Any additional circuit options left?

The measurements presented in chapter 5 clearly show the advantages of the new BestPentode circuitry. But there is at least one more option for using pentodes in small signal amplifiers: connecting the pentode in triode mode and cascoding this triode with any device you want. If we use again a npn-transistor for cascoding this circuit looks like Fig. 21.

The gain of this cascode should be in the same range like the one for the BestPentode connection. The power consumption of the whole circuit will be nearly the same too.

But, now the sum of all currents of the pentode $- I_{g2} + I_{g3} + I_a -$ is flowing through the cascoding transistor Q1. That is a major difference to the BestPentode circuit, where the screen grid current I_{g2} is flowing through the transistor only and the pentode is still acting like a pentode. Keeping in mind that

a) high-gm tubes are most useful for low-noise amplifiers and

b) these high-gm tubes need anode currents

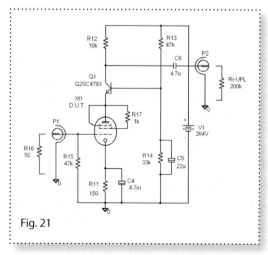

Fig. 21

of >7mA, typical up to 25mA and

c) the collector-emitter voltage is typical in the range of 50...130V, the power consumption of the cascoding transistor is in the range of >0,5W, typical up to 3W. That requires a much more powerful transistor than inside the BestPentode circuitry, where the anode is taking the hard work and not the added transistor. That has some practical consequences: First it's not possible to use cheap low power high voltage transistors like MPSA44 or other lower noise low power transistors. Instead of that for the cascode circuit special high voltage medium power transistors are necessary which cost much more money and have typical less current gain ratio. Additionally a heat sink is necessary. Because of the high bandwidth of the circuit the cascoding transistor should be mounted as short as possible to the tube – that is usually the hottest place inside the amplifier. Using an additional heatsink is not a good idea to say the least.

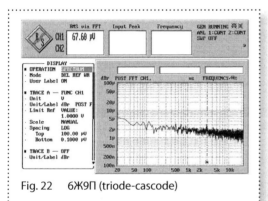

Fig. 22   6Ж9П (triode-cascode)

What about the noise of this cascode? I measured some pentodes in the cascode circuit of Fig. 21 as well. The equivalent input noise density is a bit higher than the measured values of the BestPentode circuit. As an example Fig. 22 shows the noise spectrum of the 6Ж9П in triode mode and cascoded like Fig. 21:

The measured output noise voltage of 67,60µV is a bit higher than that of the BestPentode circuitry of 65,44µV. But the difference is not much. The measured gain is nearly the same.

The measurements of other pentodes show the same results: there is no advantage in terms of noise or gain compared with the BestPentode circuit. But the much more power consumption of the cascoding transistor is a major disadvantage.

Conclusions

The investigations and measurements have validated the advantages of the new BestPentode circuit: the noise as well as the gain is improved. Furthermore the measured noise is lower than that same pentode connected in triode mode. The comparison with high-gm triodes, which have the same frame grid construction, show the potential of this new circuit.

Besides the circuit structure discussion one very well known fact is obvious: the choice of the right tube for a given application is most important. But once the right tube is chosen, the BestPentode circuit will excel in terms of noise and gain – if this tube is a pentode. The new circuit is very easy to implement in practise, just one low power npn transistor like MPSA44 is necessary. The filter network at the base could be made of much higher impedance because of the very low base current.

Consequently rather low capacitance values could be used for filtering the screen grid voltage compared with the classic pentode approach.

Following my measurement results the EF86 is the clear looser in any circuitry – there are many better options. And besides the classic circuitry using the very expensive NOS ECC808 or ECC813S, there are better – and much cheaper – options. The very good noise figures of the D3a will not surprise the experts. But the 6Ж9П is a major sleeper: it has the second best noise values, high gain in BestPentode connection and is available for less than 1 \$. And it has a very promising 1/f-noise characteristic which is a secret tip for RIAA amplifiers. Connected in BestPentode circuitry the 6Ж9П has a much lower input capacitance compared with a triode as well. That makes the adaptation of a given headshell to the RIAA amplifier much more easy. One should remember that the equivalent input noise density of the 6Ж9П in BestPentode circuitry is significantly less than that of the famous OPA627 operational amplifier!

But, please take note again that there is no statistical evidence for the listed and measured tubes. If you look at the measured EC806S you will see that both measured tubes differ very much in terms of noise. Probably the EC806S from Siemens was a bad example. That has to be proved by much more measurements. But one can get an indication where the effort is well spent for further investigations.

Maybe my investigations are a trigger to have a second look on the many pentodes which sleep in many forgotten boxes. Especially the high-gm frame grid pentodes like E180F, E186F and the like were made in exceptional good quality. During their manufacturing time they were very expensive and nobody had the idea to use them for amplification of audio signals. Today they are available in large quantities for very little money. Instead of buying extremely expensive hyped tubes like ECC813S or Telefunken made EF86

Power supply hints

High-gm pentodes like E180F and the like were designed for broadband, high frequency amplifiers. Hum suppression was not an issue for these applications. Because of that these tubes have no twisted heater wires inside the catode like typical audio tubes (EF86, ECC83). So DC-heating is a must for high-gm pentodes and triodes in low level audio devices.

There are many more promising pentodes for low-noise applications, for example: 6AC7, C3g, EFF51, PCF201 (Pentode section), PCF801/803 (Pentode section), 6Ж10П, 6Ж52П. Very important for all experiments is the use of low noise, regulated power supply and DC-heating. During the tube age such a low noise high voltage power supply needed an extremely high effort. As an example the reference voltage sources were noisy (gas filled stabilizer tube) and the regulator tube wasted a lot of additional power. Today we have easy to implement high voltage regulator circuits based on semiconductors like LM317 and we could use DC-heating with little additional effort. That opens the door to circuits and applications which were not considered in the heydays of tubes.

a fraction of this money could be spent for the very well regulated power supply – which is a must for any pentode and cascode preamplifier.

During ETF2009 one line level preamplifier as well as one PP-chokeloaded PP-power amplifier were demonstrated which have the BestPentode circuitry for voltage amplification and for the PP-driver stage respectively. The circuits of these amplifiers need an in-depth discussion and might appear in a future issue of Linear Audio.

The BestPentode circuitry is patent pending (DE 102008017678). For non-commercial and DIY use please feel free to make your own experiments with this circuitry. For commercial use please ask the author for permission.

{1} Morgan Jones, "Valve Amplifiers", 2003, p. 94

{2} Philips Datasheet E180F, 1962

{3} Philips Datasheet EF86, 1970

{4} Telefunken Datasheet ECC808, 1963

The Mini - Simplex

Ari Polisois

Recently I had the opportunity to measure the power output of several amplifiers at different sound loudness, in the same listening room, thanks to an instrument I had purchased several years ago in a Radio Shack shop. I was surprised to find that a low power output, such as 2W, produced a sound that filled the room with enough energy, except, maybe, for big classic orchestras. At loud passages, however, it started to be distorted, because the amplifier could not make it. From that, I understood that for leveled soft music without too many peak shots, a couple of watts could be enough. After some more tests, I concluded that a power output of 6-7 watts could suit many situations, and could even be, sometimes, in excess of the needs.

Some years before, I had designed a 6C33 single ended amplifier that I named the Simplex, delivering 12W at 5% total harmonic distortion, with the second harmonic strongly dominating. The latter, as you know, is more acceptable to the ears than odd harmonics. Some audiophiles do not accept single ended circuitry, because its consumption in relation to the musical power delivered is very high. Most high power valves, irrespective of the layout, spend several watts just for their heaters, plus, of course, additional power for their anode requirements.

In the 6C33's case is this amounts to:

*   Two heaters in the same bulb, each 6,3V – 3,3 A with a total of 40,6W per channel (81,2W for both channels).
*   High-voltage power supply voltage is 230-240 V and each channel absorbs at idle an average of 220 mA. This corresponds to 235V x 0,22 = 51,7 W.

The lump total would be 51,7 + 40,6 = 92,3 W. To this figure we should add about 25W for the driver stage (both channels) requirements.

I have assumed that there are no thermionic valves rectifiers, but solid state diode bridges, instead. Therefore, if we used the amplifier with 2x12W output, we would have an efficiency of 24/210 = 11,4 % . What if we used only 6W in our listening room? Obviously, the efficiency would drop to less than 6%. It's like heating a room in winter with the windows wide open. My friend Jim would say, in similar circumstances: "So what ?!" , and I agree. Spending a kilowatt to listen to your preferred music for 4 hours is not really a big sin; we could make better savings by changing some of our questionable habits. If I am not mistaken, I pay less than 20 cts for one kilowatt/hour. Nevertheless,

I decided to try some energy savings while designing an amplifier delivering 6-7W output power. I had also in mind to offer a design that would not scare newcomers interested in building their first or second valve amplifier (1).

The Simplex had already achieved the goal of significantly reducing the number of components and, as a consequence, the probability of making mistakes, and, at the same time, ensure a reduction of the cost (both in building and time spent in finding eventual defects). Now, the Mini-simplex is something still easier to understand and build. On the other hand, even skilled audiophiles cannot afford infinite time to build an audio amplifier or similar devices, because they are almost fully absorbed by their work and career and, of course, their family needs. Therefore a good amplifier project that would require just two or three available week-ends could be acceptable. The next question would be: "why not a kit?". Sure, with kits, everything is made easy, eventual mistakes avoided and the time required reduced to a minimum. However, they do not offer the same satisfaction as building your own amp from the start, see it grow step by step and, finally, listen to the results of your efforts. During this lapse of time, you have added something to your knowledge and feel more experienced, ready for the next challenge (Hear! Hear! – ed.). Finally, you can tell your friends "I have built this amp, following a project that I selected among others" and not just: "I assembled (read: "soldered") this kit, following the manual's instructions".

Valves or solid state?

This is another important choice to make.

Among the audiophiles with whom I correspond, there are many young engineers or technicians that are familiar with the solid state devices and practically ignore everything about valves. I belong to the old generation so I am at the opposite end. Maybe it is not so, but I think that nowadays the university courses dedicate very little time to explain how the valves work. It is just a small niche that does not deserve more than an hour's lesson in the whole course. My own experience is that most solid state devices cannot be easily repaired if they have a problem, except by skilled people, that sometimes are at a loss to make any progress without the factory schematics and troubleshooting manuals.

On the contrary, to find a malfunction in a valve amplifier, for instance, is much simpler as long as you have a basic knowledge of the principles, and this does not take years of dedicated studies (2). A couple of measurements, the valve's data sheet (available in seconds from the web) plus some logic, and the fault is located. With limited expense, the repair is done and everything starts working again. A few weeks ago my computer had some problems and I called a shop I had contacted before to upgrade or repair my former PCs. Because the present problem concerned a portable computer, he addressed me to someone else, adding, for my guidance, that just the initial inspection would cost me 100 $!

I gave up and because my sick computer was 4 years "old", I decided to buy a new one. The number of TV sets, audio amplifiers and other products based on transistors that I have thrown in the

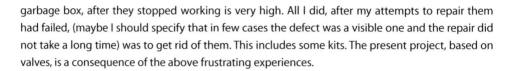

garbage box, after they stopped working is very high. All I did, after my attempts to repair them had failed, (maybe I should specify that in few cases the defect was a visible one and the repair did not take a long time) was to get rid of them. This includes some kits. The present project, based on valves, is a consequence of the above frustrating experiences.

Preliminary analysis.

I took my old Simplex unit and had a good look at it. It's the standard circuit with the Direct Coupling Modulated Bias (DCMB) structure (3). The output stage uses one 6C33 and the driver uses the two triodes of a 6SN7GT. The latter stage's circuit employs a 10k power resistor rated 50W, whose task is to supply some extra current to the driver tube cathode resistor (4). The actual consumption of this resistor alone is approx. 7W per channel, hence 14W for both. As you may have noticed, my investigation was focused on the energy budget. Then I checked the 6C33C-B data sheet, that gives the grid curves of the valve with just one 6,3 V – 3,3 A filament in use and with both (usually the two 6,3 V – 3,3A filaments are connected in series and to a 12,6 V source; see characteristic curves at the end).

From the data sheet I learned that the valve, with just one filament heated, could perform in a satisfactory way (with regard to the amplifier I had in mind), but with the limitation, however, that the anode power dissipation should not exceed 45 W (whereas, with both filaments heated, this would reach 60W).

In general, you can get, in a Single Ended topology max. 20 % of musical power from that anode dissipation static power figure of 45W, which means an average of 9W, in Class A (5). So I concluded that was a way to design an amplifier that would deliver enough sound energy and burn less watts. In fact, heating just one of the two filaments represented a saving of 40 W for both channels, which is not trivial. At first I had planned to use the same driver topology as I used before, but, in the present

TABLE #1

THE MINI-SIMPLEX - TEST POINTS VOLTAGES AND CURRENTS
BASED ON AN OUTPUT OF 6 W (6,93Vrms on a 8 ohm load)

VOLTAGES (DC except input)

A) Power supply voltage measured	276	V
B) Input	1	Vrms
C) Anode of V1 to ground	75	V
D) Cathode of V1 to ground	1,58	V
E) Grid of V2 to ground	75	V
F) Anode of V2 to ground	256	V
G) Anode of V2 to cathode	142	V
H) Cathode of V2 to ground	114	V
I) Grid of V2 to cathode	-39	V
J) B+ to anode of V2	20	V

CURRENTS

K) Anode current of V1	4,1	mA
L) Through Rx (aux.cathode current)	11,7	mA
M) Through the cathode resistor of V1	15,8	mA
N) Through the primary of the OPT	237	mA

POWER DISSIPATION

O) Rx	3,0	Watts
P) The 47k anode load	0,8	Watts
Q) The cathode resistor of V1	0,02	Watts
R) The primary of the OPT	4,7	Watts
S) The 470 ohm cathode resistor of V2	26,4	Watts
T) The 1Ω ohm cathode resistor of V2	0,6	Watts

Total consumption of the amplifier (excluding valves)	35,5	Watts

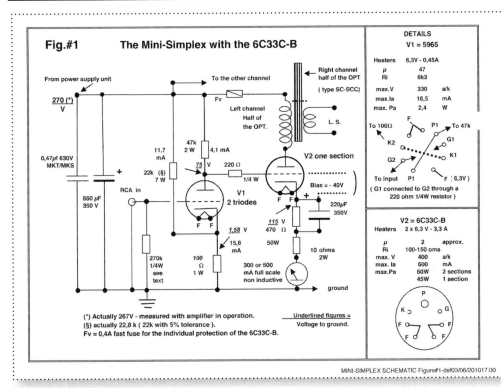

MINI-SIMPLEX SCHEMATIC Figure#1-def03/06/201017:00

instance, I was strongly tempted to try another solution investigated a couple of months before, that had proven to work properly. Its dominant characteristic is extreme simplicity, as you will see. Moreover, it is quite common and easy to understand. What more could be done? Most of the goals I was aiming for were in reach.

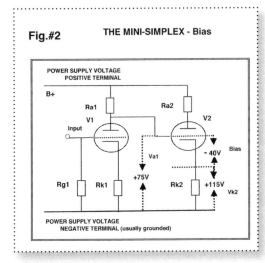

The schematic.

Fig.#1 shows the amplifier section.

For those who have some experience in valve amplifiers, understanding how it works requires little or no effort. However, for the newcomers, more explanations will be given. The driver section has been equipped with an uncommon valve, the 5965 GE, that I felt was suitable for the purpose. Its characteristics are shown in the data sheet at the end. The twin triodes are connected in parallel and we'll see why. The anodes are directly coupled to the grid of the 6C33, through a small value resistor that pre-

TABLE #2 - THE MINI SIMPLEX - PARTS LIST AND SOURCES
(Quantities are for both channels)

Item	Quantity	Description	Value	Specifications or location	⚓	Unit €	Total €	Source
37	1	Aluminum chassis	suggested 25x40cm	see text	1	40,00	40,00	*
25	2	Audio Volume controls	47k	Log	1	3,00	6,00	*
10	2	Capacitors	660 µF / 350V	electrolytic	1	35,00	70,00	*
11	2	Capacitors	0,47 µF / 630V	MKT or MKS	1	3,00	6,00	*
12	2	Capacitors	220 µF / 350V	electrolytic	1	12,00	24,00	*
47	1	Capacitors	220 µF -/ 350V	electrolytic	2	12,00	12,00	*
48	1	Capacitors	0,47 µF / 630V	MKT or MKS	2	3,00	3,00	*
34,1		enclosure - Output transformer						
34	1	Enclosure (box)	plastic or alum.	main chassis	1	12,00	12,00	*
31	1	Fuse	1,6 A	Slow	2	0,50	0,50	*
40	1	Fuse	3,15 A slow	Power supply box	2	0,50	0,50	*
13	2	Fuse	400mA	Fast	1	0,50	1,00	*
32	1	Fuse holder	250V	Standard	2	2,00	2,00	*
41	1	Fuse holder	standard	Power supply box	2	2,00	2,00	*
14	2	Fuse holder	250V	Standard	1	2,00	4,00	*
20	2	Heat sinks	2N3055 type	abt.10x6,5cm.	1	3,00	6,00	*
36	1	Input transformer	2 channels 1 inverted	see text	0	45,00	45,00	#
26	2	Knobs - for volume controls	---	Graduated	1	2,00	4,00	*
28	2	L.S. connectors BLACK	---	main Chassis	1	1,50	3,00	*
27	2	L.S. connectors RED	---	main Chassis	1	1,50	3,00	*
29	1	Mains input socket	Bulgin (computer type)	main chassis	1	2,50	2,50	*
44	1	Mains input socket	Bulgin (computer type)	Power supply box	2	2,50	2,50	*
19	2	Millammeters	0-300 or 0-500 F.S.	Non inductive	1	6,00	12,00	*
35	1	Output transformer	bi-channel type	see text	1	175,00	175,00	#
33	1	Pilot lamp - green	220V	neon	1	1,50	1,50	*
43	1	Pilot lamp - red	220V	neon	2	1,50	1,50	*
24	2	RCA connectors BLACK	for input from CDP	main chassis	1	1,80	3,60	*
23	2	RCA connectors RED	for input from CDP	main chassis	1	1,80	3,60	*
45	1	Rectifyier bridge	400V-25A	KB2504 or similar	2	4,00	4,00	*
1	2	Resistors	270k	1/4 W	1	0,10	0,20	*
2	2	Resistors	22k	7W	1	1,50	3,00	*
3	2	Resistors	47k	2W	1	0,20	0,40	*
4	2	Resistors	100 ohms	1W	1	0,20	0,40	*
5	4	Resistors	220 ohms	1/4W	1	0,10	0,40	*
6	2	Resistors	470 ohms	50W	1	6,00	12,00	*
7	2	Resistors	10 ohms	2W	1	0,20	0,40	*
8	1	Resistors	100k	10W	1	4,30	4,30	*
9	1	Resistors	33k - 7W	7W	1	1,50	1,50	*
46	1	Resistors	47 or 68 ohms 25W	wire wound	2	5,00	5,00	*
49	1	Resistors	33k - 7W	wire wound	2	1,50	1,50	*
51	1	Safety connector BLACK	1 kV insulation female	power supply box	2	2,00	2,00	*
22	2	Safety connector BLACK	1 kV insulation female	main chassis	1	2,00	4,00	*
50	1	Safety connector RED	1 kV insulation female	power supply box	2	2,00	2,00	*
21	2	Safety connector RED	1 kV insulation female	main chassis	1	2,00	4,00	*
16	2	Sockets	Noval	Ceramic	1	1,50	3,00	*
18	2	Sockets	7-pin	Ceramic	1	5,00	10,00	*
30	1	Switch	250V 4A	main chassis	2	2,00	2,00	*
42	1	Switch	250V - 4A	Power supply box	2	2,00	2,00	*
38	1	Transformer for the heaters	2x6,3V - 5A ea.	75VA	1	40,00	40,00	§
39	1	Transformer for the power stage	mains voltage to 230V	min.150VA	2	55,00	55,00	§
52	1	Twin wire Cord (mains type)	(with red and black	PS box to chassis	0	8,00	8,00	*
52,1	1	Twin wire Cord (mains type)	safety connectors - male	connection	0			
15	2	Valves	5965	Twin triodes	1	5,00	10,00	¤
17	2	Valves	6C33C-B	Power triode	1	45,00	90,00	¤
						TOTAL €	621,30	

+ Miscellaneous (wires, bolts, etc.)   ⚓ 1 = main chassis 2 = Power supply box 0 = other

SOURCES : * most electronic parts suppliers $ also Amplimo Holland   # A2Belectronic@wanadoo.fr
¤ Billington UK   Chelmer UK   Tubesandmore etc.

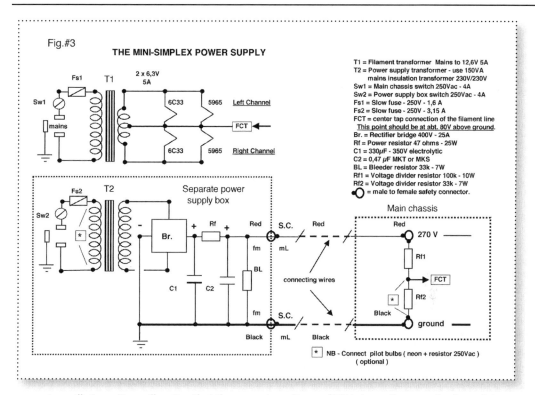

Fig.#3

THE MINI-SIMPLEX POWER SUPPLY

T1 = Filament transformer Mains to 12,6V 5A
T2 = Power supply transformer - use 150VA
       mains insulation transformer 230V/230V
Sw1 = Main chassis switch 250Vac - 4A
Sw2 = Power supply box switch 250Vac - 4A
Fs1 = Slow fuse - 250V - 1,6 A
Fs2 = Slow fuse - 250V - 3,15 A
FCT = center tap connection of the filament line
       This point should be at abt. 80V above ground.
Br. = Rectifier bridge 400V - 25A
Rf = Power resistor 47 ohms - 25W
C1 = 330µF - 350V electrolytic
C2 = 0,47 µF MKT or MKS
BL = Bleeder resistor 33k - 7W
Rf1 = Voltage divider resistor 100k - 10W
Rf2 = Voltage divider resistor 33k - 7W
○ = male to female safety connector.

NB - Connect pilot bulbs (neon + resistor 250Vac)
(optional)

vents oscillations. You will notice that they are at a voltage of 75V above the negative line of the power supply, which is grounded to the metal chassis. This requires to lift the cathode's potential to ground by 40-45V (the approximate value required for the bias) above the 5965's anode voltage of 75V. By doing so, the cathode of the 6C33 would be at approximately 115-120V above ground and a suitable biasing voltage would result between the cathode and the 6C33's grid. See Fig.#2.

A fuse has been added between the primary of the output transformer and the anode of the 6C33. It's a 400mA fast action fuse that will save the valve in case the anode current exceeds the above value. The manufacturers' specifications state that this valve can withstand up to 600mA, so we have 200mA tolerance.

The 470 ohm resistor, whose task is to raise the cathode voltage, as explained in the former paragraphs, must be placed on a heat sink and where the heat can escape freely. It becomes quite hot but not excessively, if the heat sink is big enough. I have used 6,5 x 10 cm long heat sinks normally intended for a 2N3055 transistor and have fixed them underneath the deck of the chassis, at the end sides. A look at the photo of the bottom of the amplifier will show their place.

The above resistor is by-passed by a 220µF-350V capacitor. This is necessary, otherwise the gain of the power stage would be less than unity and, what is much more important, the driver voltage swing would have to be over two times bigger in amplitude (6). In series with this power resistor you will find a 10 ohm/2W one, not by-passed by a capacitor and, therefore, producing some cur-

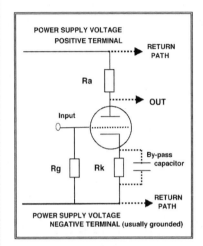

Fig.#4   THE MINI-SIMPLEX - Common cathode circuit operation

PARAMETERS

Rg = Grid leak resistor

Ra = Anode load (reistor, choke, transformer or other complex devices)

Rk = Cathode load (usually a resistor) by-passed by a capacitor or not.

Ia = Anode corrent

Ig = Grid current

Ik = Cathode current (usually the same as Ia)

Ri = Internal resistance (or plate resistance) of the valve

-Vg = Negative voltage between grid and cathode (bias).

μ = max. gain of the valve (theoretical)

AV = effective gain of th valve, without feed back

Afdb = effective gain of the valve, with feed back

Ck = Cathode by-pass capacitor.

NOTES :-

_The layout shown is named "The common cathode circuit" .

_ There are limits not to exceed regarding Ia and Va

_Normally (except with some powrfull transmitting tubes) the grid must be kept negative with respect to cathode, otherwise Ia would be too high.

_Rk is common to the grid and anode, as far as the path to the negative is concerned. As a consequence there is an interaction involving both these electrodes.:-

_The anode current Ia produces a voltage drop between Rk's terminals.

_ This voltage drop is positive at the cathode and negative at ground level.

_The grid finds itself at the same voltage, but reversed. This is called the bias and it sets the operation point at still condition, of the valve.

The data sheet of the valves shows the variation of the anode current Ia at various bias levels, that are shown next to each curve.

_With regard to an altrnating current signal, the fact that Rk is by-passed by a capacitor or not, makes a difference in the effective gain of the valve.

SEE FORMULAE #1 AND #2 ;

FORMULA #1 - Case of unbypassed Rk

$AV = \mu * (Ra / (Ra + Ri)$

FORMULA #2 - Case of by-passed Rk (where the reactance of Ck is very low (= high capacity) compared to Rk

$Afdb = AV / [1 + (\beta * AV)]$ Where $\beta = Rk / Ra$

REMARK : If you replace the cathode resistor of 100 ohms (Fig.#1) by a 387 ohm one, the effective amplification factor drops from 38 to 31.

Valve type 5965 $\mu = 47$ Ri = 6300 ohms   Ra = 47000 ohms

---

rent negative feedback (local feedback in this case because it acts between electrodes of the same valve). This local feedback will have three main consequences:-

1. the distortion will be reduced;
2. the gain will be slightly affected (as it depends on : the ratio between impedance of the load – our output transformer has about 540 ohms – and the value of this resistor – 10 ohms – as well as on the amplification factor of the valve (say about 2);
3. It raises the voltage of the cathode to ground by 10 ohms x 0.23 A (the latter being the average idle current of the 6C33), that is 2.3 V.

This is acceptable. There is also a milliamp meter in the negative power supply line to monitor the anode current of the 6C33. This is not critical and can be deleted without consequences, except that it is always interesting to check that the valves operate at the design current, which happens to be, in our case, approx. 230mA per channel (7).

The driver stage.

Nothing really odd. The valve (5965, with the twin triodes paralleled), its anode load resistor (47k), its cathode resistor of 100 ohms and its grid leak resistor of 270k that, in the unit I have built, has been

replaced by an audio (Log) volume control, make a familiar and recurrent circuit. But there is an additional resistor of 22k, which supplies some additional current from the supply to the cathode. This current is about 11,7mA, against the 4,1mA of the paralleled triodes. Why did I add this resistor? Very simple. First of all, I had calculated that the amplifier should deliver about 6W with 1V RMS at the input, which corresponds to 1.4 V peak.

This means that the bias of the driver triodes must be somewhat above this level, to avoid the grid current that causes distortion. For the above reason, the bias for the 5965 must be, say, 1.5-1.6 Vdc. In our circuit, this bias is ensured by the presence of the cathode resistor. Knowing the anode current (either by calculating it with the data sheet, or by measuring it) which is about 4mA, this would force us to use a 1,55V / 0.004A = 387 ohm resistor. If you try this solution, forget about driving the amp to full power with just one Volt rms. The local negative feed back that will take place (I repeat: this depends on the ratio Ra/Rk – where Ra is the anode load and Rk the cathode resistor) will reduce the gain of the 5965 considerably and you will have to change something (valve type, power supply voltage, anode load, etc.). This is not impossible but I prefer to use the additional 22k resistor. Moreover, there is something else to justify this choice (see Fig.#4).

The additional resistor connects directly the positive line of the power supply to the cathode of the driver valve. What happens is that any residual noise or hum, still present after filtering, will be sent to the cathode which, in turn sends the same, reversed, to the grid. As a result, there is a subtraction of noise and hum up to 75% of the original amount. It's a Canadian friend of mine, Tom Clearwater, who called my attention to this fact. I tested it and found it was true. So, why not adding this affordable resistor and have all the mentioned advantages?

Next, a few words on the choice of paralleling the twin triodes of the 5965.

Although the amplification factor (μ) does not change, the resulting internal resistance is halved, and this is an improvement that keeps the high frequency range of the driver stage as wide as possible. Because of the high anode to grid capacity of the 6C33 and the Miller effect, the lower the value of the resistor present between its grid and cathode, the better for the frequency range. In simple words, it's like discharging a capacitor. The lower the resistance connected to it, the faster this happens.

The power supply.

Standard circuit, standard components. I always suggest to build this stage first and also to keep it separate from the main chassis, whose sensitive components can be affected by the hum spread around by the power transformers and, eventually, the chokes. It helps building this unit first, testing and approving it, before proceeding with the more difficult part, the amplifier. To simulate its behavior in operation (with the expected current flowing) you can use, for instance, several resistors in series or parallel, to get about 600 ohms, provided the resistors can handle about 200W (actually 125W = 270V x 0,46A) in total.

The transformer that supplies the voltage to the heaters is less critical and can be placed on or inside the main amplifier chassis. It is not a big transformer (2 x 6,3V – 5 A = 65VA of power or, better, 75VA) and it is preferably located at the opposite end to the output transformer. If you want my advice, use a large chassis. Mine measures 40x21 cm, but a 40x25 is preferred. The separate power supply box comprises the power transformer, the rectifier bridge, the reservoir capacitor and a filtering resistor, plus, of course the switch, fuse and a pilot lamp. Moreover, I fitted a 33k – 7W resistor after the filtering resistor, to discharge the reservoir capacitor. Some capacitors stay charged for a long time after you switch the mains off and this might be dangerous. With the above bleeder resistor, it takes twenty seconds or less to discharge. The connectors are of the safety type, namely all terminals are insulated and do not protrude bare. A simple two-wire power cord connects the power supply box to the main chassis, where safety connections of the same kind are used. The length of the cord is not critical at all. The power supply box switch must not be actuated at least 10 minutes after the filament transformer is switched ON. This gives the valves enough time to warm up before getting the main juice. By the way, you will notice that the sound will improve after some more time, estimated in another 20 minutes. This is a well known characteristic of some power valves, including the 6C33s.

The output transformer.

I selected the SC-SCC (Split Core Stereo Common Circuit) output transformer because it is the best for such a compact amplifier. In fact, this transformer is less bulky than any other kind, as the two channels share the same magnetic circuit (8). This transformer can deliver a lot of bass, as you can see from the graph of the frequency range. Moreover, because it requires less primary turns, it also ensures an extended high frequency range. The channel separation is satisfactory and the sound

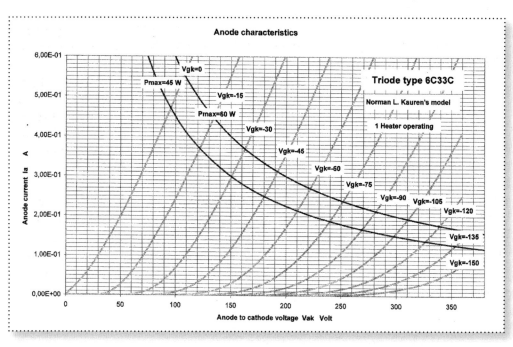

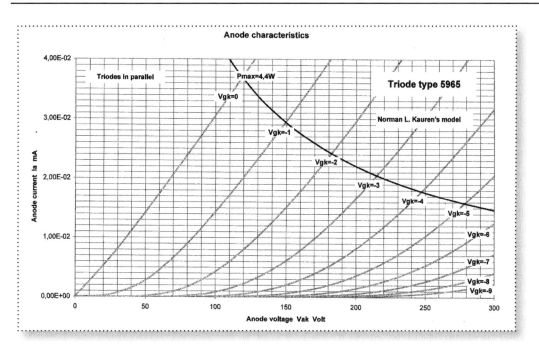

Anode characteristics

fills the room with strong bass and clear midrange and treble waves. It's a real good sensation.

The input transformers box.

This unit consists of two input transformers with the ratio 1:1, one of which is non inverting and the other inverting. A phase inversion is necessary when using the SC-SCC output transformer. More information can be found at the ETF 2008 site, under "lectures by Ari Polisois".

The frequency range of this unit is very large (from few Hz to over 200 kHz) and the input and output impedances adequately high. Please refer to the 3B input transformer set description that you find in the "Products" section of my site www.polisois-audio.com . (For availability of the Split Core Stereo Common Circuit transformer contact the author at A2Belectronic@wanadoo.fr or see www. polisois-audio.com - ed.)

Conclusion.

Nothing is purely simple or purely easy, but you must admit that the comparison between the mini-Simplex and many other amplifiers in the 6 W range speaks for itself. Wait until you and your friends and relatives listen to the Mini-simplex' voice and I am sure you will agree on its qualities. Before that, never forget that valves operate with high voltages, unlike transistors (this I must admit), so always be very careful when working on this amp. If you need some additional explanation, you can contact me at: A2Belectronic@wanadoo.fr or www.polisois-audio.com

Footnotes.

1. The circuit I have systematically adopted has been the Direct Coupling Modulated Bias, that ensures noticeable improvements in the sound quality, but I have noticed that only few audiophiles have really understood its advantages or, if they did, they consider it tricky and complex, which is absolutely not the case. Some described it "unstable". On the contrary, stability is one of the basic features of this circuitry. Nevertheless, beginners should be treated with careful attention, and it is strongly advisable not to discourage them from the start. They will have more opportunities, later on, to understand the DCMB merits.

2. Good books, to start with are:
   - Valve amplifiers (Morgan Jones – Newnes UK)
   - Modern valve amplifiers from 10 to 100W, Menno van der Veen – Publitronic/Elektor
   - Audio and HI-FI Engineer's Pocket book (Vivian Capel – Newnes UK).
   - An article of a well explained project, combining theory and practice, also helps a lot.

3. More on this subject on my site www.polisois-audio.com

4. This resistor helps keeping the cathode bias resistor at a low value. Because the latter is not bypassed by a capacitor, a higher value would reduce the gain of the stage. The choice of the 22k resistor and Rk (the cathode unbypassed resistor) is a compromise between the gain of the stage and local negative feed back, that reduces it but also the distortion.

5. Class A means that the peak of the signal that drives the power tube grid must not be in excess of the DC bias level.

Behind the scene.

You might be interested in the itinerary of the design, just in case you wanted to try another alternative (with other valves, for instance) or understand the reason of some choices. Bear in mind that, with the SC-SCC Opt and low internal resistance valves (6C33, 6336, 6080 etc., that is, with Ri below 300 ohms) you achieve a very wide frequency range and that is important for the timbre and substance of the sound. In two words, hereafter is what I did for the Mini-simplex, for your guidance. First I took the 6C33 data sheet (the alternative with one 6,3V heater in use) and selected an operating point corresponding to a medium-low anode voltage (140-180V). At 150 V between anode and cathode, I calculated that I could not exceed 45W / 150V = 300mA (max. plate dissipation divided by the voltage). I then decided to stay a little below, namely 200 mA. I found the necessary bias corresponding to the above idle anode currents, which is approximately 40V (because the valves do not necessarily follow the theoretical curves be careful not to give 100% credit to the preliminary estimations). Considering that driver valves normally operate, in a layout similar to the one described, with an anode to ground voltage of 70 to 100 V, I found that the cathode resistor of 6C33 should be at least 70+40=110 V above ground. This voltage depends on the value of the cathode resistor and, of course, the anode current that theoretically would be 200mA. The calculation is simple: 110V / 0.2 = 550 ohms. I selected a standard 50W resistor of 470 ohms that, with the addition of the 10 ohm resistor (the one unbypassed in the schematic) would bring us to 480 ohms. Reality decided that the current would be somewhat higher and the cathode to ground voltage settled to 115 (not too far from the forecasts). Table #1 gives the measurements found after completion of the Mini-Simplex.

6. The grid to cathode swing that is necessary to get full power, is, in our case, close to 40V peak. However, part of the AC signal amplified by the 6C33 would be present at the 470 ohms resistor terminals, if it was not by-passed by the capacitor. A rough estimate would be 50 V, which would oppose the 40 Vac supplied by the driver. The latter would have to be raised to 90 Vac to restore the net swing of 40 Vac.

7. I used the expression "happens to be" because it is rather difficult to calculate, in advance, the anode current that results from several interactive factors. The B+ voltage, the cathode resistor of the 6C33, the anode resistor of the 5965, the internal resistance of the latter valve, its cathode resistor also, etc. , all of these participate to fixing the anode idle current of the power valve. That's not all; once the current stabilizes, it affects the voltage between the anode of the 5965 and its consequences, and so on… You better forget what I just said. What matters is that the 470 ohm cathode resistor and the power supply voltage are dominant and, in practice, the anode current is stable and you will find no significant difference between one listening session and the next.

8. INPI (Institute National the la Propriété Industrielle - Paris) patent 03 10898 - granted to Ari Polisois and Giovanni Mariani (Manager R&D of Graaf - Modena - Italy), following patent 01 02457 (INPI - Self compensated transformer with flux escape). Presented at the AES convention held in May 2005 in Barcelona - Spain

Split the difference -
The Truth about the Humble Cathodyne

Stuart Yaniger

Introduction

If one's goal for a tube amplifier is to have the output replicate the input to the greatest extent possible, then a push-pull output stage will be required. For a given pair of output tubes, push-pull amplification will yield the lowest distortion, the highest power output, the highest efficiency, and greatly ease the design requirements for the driver stage and the output transformer. While there may be design goals that call for single-ended amplification, input-to-output accuracy is not one of them.

The one complication of a push-pull design is that not only must a signal be delivered linearly and at high level to the output stage (the "push"), an exact mirror image of that signal must also be supplied (the "pull"). The so-called "phase splitter" is that mirror.

What is a phase splitter?

A phase splitter is, simply, an amplifier circuit that has one input and two outputs: one output with the same polarity as the input signal and one output with opposite polarity. In other words, if the input to the amplifier is f(t), the amplification (assumed linear) is A, then the first output would be Af(t) and the second output would be –Af(t). Another way to look at this circuit is that it (ideally) takes a single ended signal and outputs a balanced signal.

Why is this sort of circuit called a "phase splitter?" It doesn't split anything nor does it have any effect (beyond normal bandwidth effects) on phase. Probably for the same reason that the Holy Roman Empire was neither holy, Roman, nor an empire. Or maybe because "single ended to balanced converter" is too long to type. In any case, we bow to tradition in our terminology and will use the term "phase splitter" to mean "a circuit that converts a single-ended signal to a pair of signals, one inverted and one non-inverted with respect to the input."

One more set of terminological standards that will be used in this article: internal tube parameters will be indicated in lower case (e.g., r_p, the plate resistance; r_k, the cathode impedance), external

parameters will be referred to in upper case (e.g., R_p, the plate load resistor; R_k, the cathode resistor; Vin, the input voltage).

Desired qualities

An ideal phase splitter would have the following properties:

1. High input impedance
2. Low output impedance
3. Low distortion
4. Perfect balance
5. High bandwidth
6. Insensitivity of balance to tube variations and aging

As usual, there is no perfection in the universe, but two phase splitter circuits come close, the long-tailed pair and the cathodyne. There's a fundamental similarity between the two and some fundamental tradeoffs.

First, we review some very basic facts about triodes. We will not derive these rigorously (derivations may be found in any standard tube text), but we will make basic use of several approximations:

1. A triode may, to first order, be modeled as a voltage (Thevenin) source in series with effective plate resistance. The voltage source is equal to the tube's μ multiplied by the grid-to-cathode voltage.

2. The effective plate resistance is equal to r_p plus any resistance between cathode and ground multiplied by $(\mu + 1)$.

3. The effective impedance looking in to the cathode is the total plate resistance (load resistance plus r_p) divided by $(\mu + 1)$.

With these rules and some basic Ohm's and Kirchhoff's laws in hand, we have the tools we need to do basic circuit analysis.

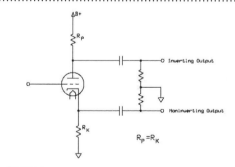

Figure 1

The cathodyne (Figure 1) is the simplest possible phase splitter- one tube, two resistors. The equivalent circuit, shown in Figure 2, tells the tale- Kirchhoff's Law forces the voltages across each of the load resistors to be equal, if we assume that the loads at each output are equal.

That last statement has caused no end of confusion- it LOOKS like the cathode output ought

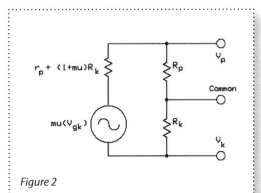

Figure 2

to have a lower source impedance than the plate output and should thus react to changes in the load impedance or the inclusion of reactances (e.g., the input capacitance of a following stage). But consider the equivalent circuit in Figure 2 where we load each output with a capacitance. One would naively expect that the frequency response at the plate to suffer more than the frequency response at the cathode. After all, at the plate, with rising frequency, the load impedance decreases, thus reducing gain. But the restriction that both outputs be loaded equally means that the $(1 + \mu)R_k$ portion of the effective plate resistance also drops, serving to increase gain at the plate.

An alternate way to look at this is to consider the plate circuit to have a larger Thevenin resistance but at the same time, to have a larger generator voltage. With changes in loading, as before, both the generator voltage and the Thevenin resistance change. Consider, for example, a decrease in the load resistances. The Thevenin resistance decreases, but because the cathode voltage is effectively in series with the input voltage, the generator voltage decreases as well. In Preisman (1960), the equations for this model of the cathodyne are rigorously and elegantly derived.

Gain

With some simple algebra, one can see that if the plate and cathode loads are equal, the output voltage is given by

$$Vout = \mu VinR/[r_p + (2 + \mu)R],$$

where R is either the cathode or plate load impedance, R_k or R_p. This expression is derived step by step in Preisman (1960) from the equivalent circuits, and separately by Jones (2003) using a feedback argument. This expression can be simplified a bit by recognizing that usually

$$(2 + \mu)R >> r_p$$

so that we can approximate the gain by

$$Vout/Vin \sim \mu/(2 + \mu).$$

This is somewhat less than unity, and that's the other common knock on the cathodyne. But let's consider things on a tube by tube basis- a long tail pair (the other type of phase splitter whose bal-

ance is forced by Kirchhoff's Law) uses two tubes (or tube sections) and achieves a gain somewhat less than half the tubes' μ. If we use a second tube with the cathodyne, configured as a common-cathode amplifier, we can do somewhat better. The cathodyne, having 100% degeneration in the cathode circuit, has a very high input impedance and will not load down the driving common cathode stage, so the combination of the two tubes can have a gain nearly equal to μ.

Source Impedance

OK, we understand that whatever the source impedances of the cathodyne's cathode and plate outputs are, they act as if they are equal under the equal-loads boundary condition. But what IS that impedance? Is it the high impedance of a plate or the low impedance of a cathode?

The Thevenin source impedance can be approximated by dividing the open-circuit voltage by the short circuit current. The former is given above; the latter can be determined by consideration of the equivalent circuit in Figure 2. Because of the equal-load boundary condition, BOTH loads need to be shorted; thus, the short circuit current is simply $\mu V_{in}/r_p$. With a bit of algebra, the source impedance from either terminal is then Zout = $Rr_p/[R(\mu + 2) + r_p]$ which, for high μ, approximates to $1/g_m$, similar to a cathode follower. The equivalent circuit for this boundary condition (equal load) is shown in Figure 3.

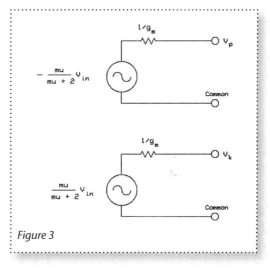

Figure 3

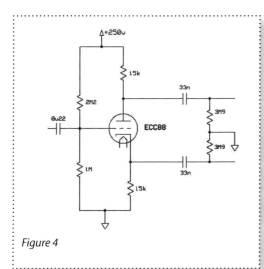

Figure 4

How good is our logic? For test purposes, we will use an ECC88/6DJ8 in a basic cathodyne circuit. This tube has been successfully used in this role many times in the past (see for example Jones 2003 and Yaniger 2009). It has good linearity, low internal impedance, and is common and inexpensive. The test circuit is shown in Figure 4. The grid resistors bias the cathode to about 1/3 of B+ in order to set the operating conditions to something close to optimal; this is often accomplished in practical amplifiers by direct coupling a preceding voltage amplifier stage to the cathodyne grid.

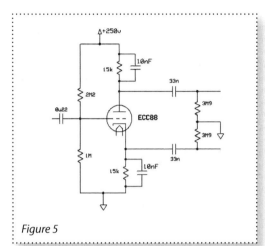

Figure 5

Let us make some predictions about the source impedance of each of the nodes of our test circuit. For the test circuit, the plate and cathode current is 5.3mA. Referring to the graphs of μ, transconductance, and plate resistance on the ECC88 datasheet, we see that at this current, g_m is about 7.2mA/V, $\mu = 31$, and $r_p = 4k6$. The approximation would then predict the looking-in impedance to be about 140R. Our earlier symmetry argument predicts the same source impedance from the plate as well.

To test this prediction, a source impedance test circuit, shown in Figure 5, was built. In this circuit, the source impedance may be determined by feeding in a square wave, then measuring the rise time. Rise time is usually defined as the response to a step function (e.g., a square wave), and specifically the time it takes the signal to rise from 10% to 90% of the full value. Rise time can then be converted to the more conventional time constant (RC) by dividing by 2.2.

Figure 6 shows the superimposed waveform response at plate (the waveform that starts high and goes low) and cathode (the waveform that starts low and goes high). First, note that the time responses are nearly perfectly symmetrical as demanded by Kirchhoff's Law. The effective source impedances at cathode and plate, and hence rise time, are equal. Second, note that the rise time (or fall time, in the case of the plate signal) is about 6 µs. That corresponds to a time constant of (6 µs)/2.2 = 2.7 us. Working backward R = t/C = (2.7 µs)/0.01 µF = 270R, in general agreement with predictions- the source impedance is low. It is left as an exercise for the reader to explain the factor of two difference between the model and the individually measured rise times (hint: symmetry!)

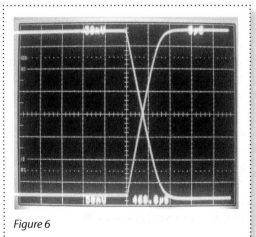

Figure 6

This is very important to note, yet again: with both terminals loaded equally, the effective source impedance at each terminal is equal. Because of the symmetry, the plate circuit can drive a capacitance equally as well as the cathode circuit. The equivalent circuit of Figure 3 is validated.

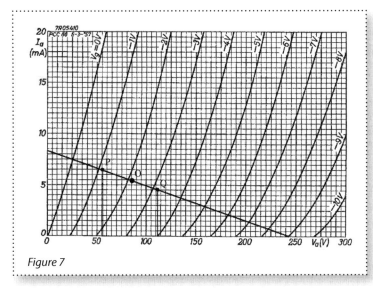

Figure 7

Distortion Performance

As may be expected from a circuit with high local feedback, the distortion performance of a catho-dyne can be quite good. As with any other triode amplifier design, it may be appreciated that dis-tortion decreases with in-creasing load impedance. And as with any other triode design, the use of load lines can be predic-tive and useful.

Let's examine the operating conditions of our test circuit, then see how it actually fares driving ca-pacitive and resistive loads. Figure 7 shows the plate characteristics of an ECC88. We note that the total load is the sum of the plate and cathode resistances, or 30k. If the tube were operated as a common cathode amplifier, the load line would run from the 250V B+ to $(250V/rp) = (250/30000) = 8.3mA$. That load line is drawn on the plate curves, with the quiescent point indicated by the letter O.

The gain of this circuit with no cathode degeneration can be calculated from

$$A= -\mu Rp/(Rp + rp) = (31)(30k)/(34k6) = 27.$$

We can see that a grid-to-cathode swing of approximately 1V yields an output swing of approxi-mately 27V, and this is indicated by the points marked P and Q on the loadline. We can also see that, within the eyeballing approximations, the distortion is fairly low to begin with and we are well within the swing capability of this circuit. In fact, it appears that this circuit could easily swing +/-65V without danger of clipping or gross distortion.

Jiggering this circuit to make a cathodyne, we split the load equally between cathode and plate. Therefore, the swing at each terminal is dropped by a factor of two (though the terminal-to-terminal swing is constant). The resulting +/-32V linear swing is adequate for many output tubes, but cer-tainly not for all- driving lower sensitivity output tubes should be accomplished by using a higher voltage cathodyne or interposing another stage of voltage amplification between the cathodyne and the output stage (e.g., the classic Williamson amplifier described in Jones 2003).

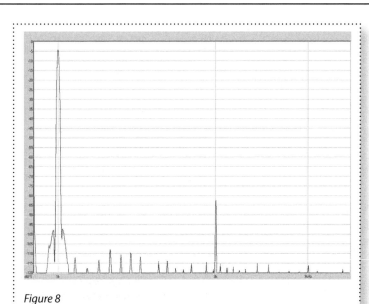

Figure 8

Let's see how this translates to the real world. The circuit of Figure 5 was set up and fed a 5VRMS sine wave. The plate and cathode outputs were connected to opposite phases of a balanced input to a sound card. This loaded the cathodyne with 10k and about 180pF between cathode and plate. The distortion spectrum is shown in Figure 8. The dominant second order distortion (-79dB, about 0.01%) is a result of imperfect matching of impedances- with tighter matching of plate and cathode loads, this could be nulled, if desired. Third harmonic was down over 110dB, and 4th and higher were unmeasurable.

Adding capacitors in parallel with each of the loads (in this case, 10nF) gave the spectrum of Figure 9. The distortion is slightly degraded, with third order significantly increasing and fifth order popping out of the noise, but it's still very, very low.

This fine balance, distortion, and drive capability is achieved with a very common and inexpensive tube and a pair of matched resistors.

"Fixing" the "Mistakes"

Because of widespread misunderstanding of the fundamental properties of the cathodyne, many "fixes" have been suggested for its nonexistent problems. Prime among

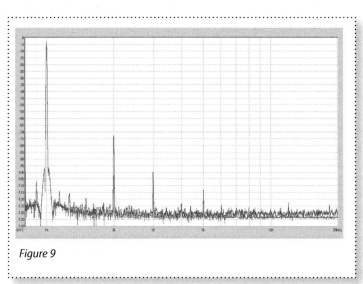

Figure 9

the nonexistent problems is the nonexistent differing source impedances between cathode and plate outputs. Many textbooks and magazine articles have suggested that these nonexistent imbalances be addressed by hobbling the cathode output with a series resistor to somehow "equalize" already-equal outputs.

As one might surmise from the "not fixed" results shown in Figure 6, the outcome of this approach will not be a happy one. The usual suggestion for the build-out resistor is the difference between Rp and the source impedance at the cathode, in this case 15k – 275R, or 14k7. To test this suggestion, the circuit of Figure 10 was built. This is identical to the circuit of Figure

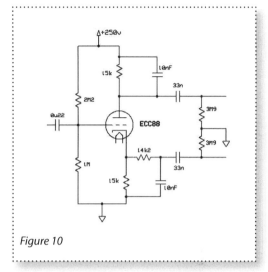

Figure 10

6, but with a 14k resistor inserted between the cathode output and the capacitive load. After reading through this analysis, it is clear that this now unbalances the cathode and anode loads, which we predict to unbalance the output amplitude and frequency response.

Figure 11 shows the bad news, with the test circuit being driven by a 1 kHz square wave. In this photo, as before, the cathode and anode waveforms have been superimposed to highlight imbalances. The plate output begins low at the left of the image; the cathode output begins high at the left. Using the recommended build-out resistor, we have now unbalanced the load on each half of the cathodyne. Putting the capacitance at arms' length from the cathode, the cathode impedance is now seen not to decrease with frequency in the same way as the plate impedance. This has severe consequences for overall circuit balance, which is visibly degraded, bandwidth (the rise time has gone from 6µs at each terminal to 200µs at the cathode and 500µs at the plate!), and frequency balance. The cure for the nonexistent disease has given the patient cancer.

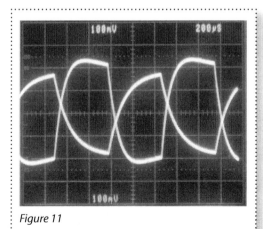

Figure 11

Another common nonexistent problem is the capacitive unbalance when driving a stage that is biased to class AB or B. The logic is that, when a tube is cut off, its amplification vanishes, thus Miller capacitance vanishes as well. Therefore, in a push-pull circuit that is biased outside of class A, the effective load on each half of the cathodyne changes as one tube cuts off and

85

the other is handling the signal. This argument is correct as it goes- however, driver stages are al-most never run in anything but class A. Output stages are a different matter. But output stages are almost invariably loaded with an output transformer, and even though one side of the push-pull output stage might be cut off, the plate voltage still changes with the signal because of the action of the transformer in coupling one side of the push-pull pair to the other. It has been suggested that cathodynes not be used with push-pull output stages in class AB or B, but that appears to be an unwarranted restriction.

Likewise, it has been suggested to "slug" the outputs of a cathodyne with small capacitors to mini-mize the effect of changes in Miller capacitance in class AB and class B output stages. This can be seen to be unnecessary.

Defects in the Argument

There is an actual source of high frequency imbalance that we swept away with our "no grid cur-rent" assumption, namely, the feedback capacitance from plate to grid, indicated as Cgp in Figure 12. If we presume that the source impedance is low, then because the plate signal is approximately equal to the opposite polarity of the input signal, there is effectively a capacitance of two times the plate-grid capacitance between the plate and ground, with no corresponding capacitance in the cathode circuit. Remember, the cathode is "following" the input signal, so the effective grid-cathode capacitance is reduced by $(\mu + 1)$ from the degeneration at the cathode.

This capacitive imbalance is pretty small (for the ECC88, less than 3pF), but for tubes where it might not be quite so small, or even MOSFETs, this effect may be compensated for by means of a small trimmer capacitor in par-allel with the cathode load resistor, as shown in Figure 12. The trimmer can then be adjusted to null out this effect and any other capacitive im-balances between plate and cathode circuits.

Another defect in the simple-minded model is rejection of noise from the power supply (PSR). If the model of Figure 3 were perfectly accurate, the PSR of each of the

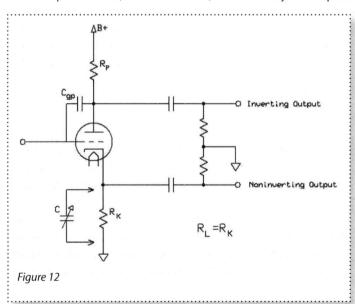

Figure 12

two outputs would be the same. In the case of the cathodyne, this is not so. This is because of the symmetry requirements- the power supply modulates the plate output directly, but the cathode output divided down by $(\mu + 1)$, so the effective load is no longer symmetrical. The power supply noise is thus not common-mode and not rejected by downstream push-pull circuits. When implementing a cathodyne, this must be considered in the noise budget.

Finally, there is lower common-mode rejection than from a "true" Thevenin source. Common-mode signals can also cause the effective load to be unbalanced since an equal but not oppo-site voltage is injected into each of the outputs, thus the rejection at the plate is significantly lower than at the cathode.

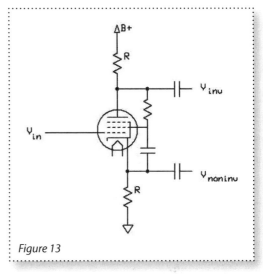

Figure 13

One More Variation

It may be appreciated that a pentode could also be pressed into cathodyne service. The complication is that the plate and cathode currents are not generally equal- the cathode current is actually the sum of both plate and screen grid currents. Further, the screen must have a fixed potential with respect to the cathode, and in the cathodyne, that means the screen must move in synchrony with the output signal. One circuit which can overcome these objections is shown in Figure 13, where the usual screen resistor is now connected to the plate, rather than B+, and where the screen is by-passed to the cathode, rather than ground. This could also be accomplished by a floating DC power supply or battery, with the negative end connected to the cathode and the positive end connected to the screen grid.

Summing Up

When loading of plate and cathode circuits is equal, the cathodyne acts to a very close approximation like two separate Thevenin sources of opposing polarities, the output of each in series with an identical (and low) impedance. The restriction that cathode and plate loads be equal means that common mode rejection is compromised. Further, the power supply rejection is likewise also compromised.

Cautions regarding capacitive loading and the effects of driving high input capacitance output stages are easily disposed of - a simple cathodyne is capable of driving relatively high capacitances from BOTH terminals with low distortion and high bandwidth. The source impedance from both terminals, when loaded equally, is comparable to a cathode follower. No special precautions need

be taken for class AB or class B output stages.

The gain of a common-cathode voltage amplifier coupled to a cathodyne is approximately double that of an LTP phase splitter. However, an LTP will have the advantage of even-order harmonic cancellation. The swing of a cathodyne from each terminal is approximately half that of the same tube connected as a common-cathode voltage amplifier.

ACKNOWLEDGEMENTS

I would like to thank Jan Didden for the use of his shiny new soap box. As always, I had lots of people teaching me, and I should single out Chris Paul, Dave Slagle, Brian Beck, and Morgan Jones for making me less stupid.

BIBLIOGRAPHY AND REFERENCES

"Notes on the Cathodyne Phase Splitter", Albert Preisman, Audio, April 1960.

"Understanding Hi Fi Circuits", Norman Crowhurst, Gernsback 1957.

"Valve Amplifiers", 3rd ed., Morgan Jones, Newnes 2003.

"Modern High End Valve Amplifiers", Menno van der Veen, Elektor 1999.

"Electronic Designers' Handbook," Robert Landee, Donovan Davis, and Albert Albrecht, McGraw-Hill 1957.

"Electronics", William Elmore and Matthew Sands, McGraw-Hill 1949.

"The ImPasse Preamplifier", Stuart Yaniger, AudioXpress, February 2009.

"The Red Light District", Stuart Yaniger, available at http://syclotron.com/?page_id=3

Datasheets E88CC, ECC88 Philips Electronics.

STEREO - From live to recorded and reproduced - What does it take?

Siegfried Linkwitz

1 - Introduction

The stereo format has been in use for a long time. I still remember how impressed I was by the spaciousness and volume of sound from a large Hi-Fi console that I heard during my time at Telefunken. I was working in the Electro Acoustic Laboratory as a "Werkstudent", in the prime days of vacuum tubes and to fulfill requirements for studying electronics at Darmstadt Technical University in Germany. My interest was not so much in audio but in higher frequencies having built my own 142 MHz ham station from WW2 components during highschool years. So I graduated with a focus on radio frequency and microwave technology and spent nearly four decades developing electronic test and measurement equipment in the laboratories of Hewlett Packard Co. in California. Audio, though, became a hobby and together with several other engineers I worked during off-hours on projects designing FM tuners, stereo decoders, preamplifiers and power amplifiers. All our designs used solid-state components, not tubes that were familiar ground. We wanted and needed to learn about the idiosyncrasies of the latest Germanium and Silicon bipolar devices, diodes, field-effect transistors and operational amplifiers while we built equipment for use at home. We loved music but were generally dissatisfied with the loudspeakers we owned. Loudspeakers challenged us, because we had difficulty understanding the reasoning behind their design. Much of that was apparently based on someone's "golden ears". We tried to measure and equalize the sound from the loudspeakers that we had purchased, having designed and constructed our own third-octave real time spectrum analyzer. HP was not in the audio test business and it was some time before an FFT based analyzer could be readily purchased. We learned.

We viewed the loudspeaker as a 20 MHz to 20 GHz broadband antenna, because a 20 Hz to 20 kHz acoustic radiator covers a similar range in wavelengths. Knowing something about antennas and their radiation patterns we wondered about the loudspeaker driver arrangements on the front baffle, about the combined output from different drive elements in space and how that is affected by the electrical drive signals used. Eventually this lead to the Linkwitz-Riley acoustic crossover filter functions [1] and greater attention has been paid to the radiation pattern of a loudspeaker ever since. We looked at the loudspeaker as a transducer, which converts electrical signals into acoustic signals. Somehow the room had to be involved in what we heard [2] because a loudspeaker radiates in many directions. My interest in loudspeaker design, loudspeaker placement in rooms, and with the difficulties of reproducing what I hear in a concert hall, began at HP in the late sixties.

The problematic with STEREO starts at the recording end of the signal chain. The sound transmission from the live event to its reproduction in a small room must be treated as a system, if realism and believability are the goal. My acoustic reference is live, unamplified sound that I pay attention to and remember. I firmly believe in the importance of acoustic and electrical measurements to correlate graphical data with what I hear and not hear. My reliance upon measurement to quantify and control phenomena comes from many years in research and development of electronic test equipment.

2 – The live experience

I have held season tickets for local symphony orchestras ever since I came to California, because I enjoy music and the arts. This has given me many opportunities to study what I hear and also to periodically refresh my memory of what live, unamplified instruments sound like. This is essential knowledge for evaluating recordings and loudspeakers. At my seat in Symphony Hall I am subject to sound streams [3] coming directly from the instruments on stage on their shortest path. Superimposed to them are reflected streams of sound arriving from many directions, from the canopy above the orchestra, the diffusing walls behind the musicians, the sidewalls and balconies and even the very high ceiling. The reflections convey a sense of space, of width and height of the hall. The reflected streams bounce around, get diffused and die out slowly when the music turns quiet or ends. The response of the hall, the reverberant sound, is an essential part of my listening experience.

Fig.1 - Symphony Hall and Orchestra as seen from my seat

Reverberation must be in a pleasing proportion and delayed to some extent when compared to the direct sound [4] at any listener location in order to preserve clarity and articulation of the musical instruments. The canopy above the orchestra has curved reflector panels. From my seat I can see all the instrument groups mirrored in the variously angled panels. They send reflected high frequency streams towards me. The panels reflect low frequency sounds less effectively, because of insufficient size and wide gaps between them. Those sounds pass through the canopy and are reflected by the larger surfaces behind them. Concert hall design is a combination of science and art to please sponsors, audiences and musicians. The halls with the largest number

of "good seats" are usually considered to be the best. Scientists attempt to understand why and define acoustic measurements, such as for reverberation time RT60, early decay time, lateral-energy fraction, direct-to-reflected sound ratio or sound energy growth, to correlate them with listener preferences.

Sound streams arrive at the ears, but hearing happens between the ears, in our brain. It is a cognitive process that takes the superimposed air pressure variations at each eardrum, the visual cues from the eyes, and how both change with head movement, to form a mental picture of the auditory scene [3]. In this picture, with eyes closed, the orchestra is at some distance in front of me and subtends a certain angle. The orchestra resides inside a space that has width, height and depths. I am aware of the nature of this space because it has responded in a unique way to the orchestra's sound. This makes it a concert hall rather than an outdoor concert experience.

3 – Binaural recording and playback
To accurately reproduce at a later time what I heard in Symphony Hall would require placing tiny microphones inside my ear canals in front of my eardrums [5]. That is not easily done and potentially harmful. Instead I have tried small omni-directional microphone capsules right next to the entrance of the ear canal. When such recording is played back over two loudspeakers I hear a high frequency coloration of sounds. The sound spectrum had been modified by reflections off the pinna, the outward part of the ear, during recording. While this is a natural phenomenon that we use for directional hearing, it becomes unnatural when the modified spectrum coming from the loudspeakers is changed a second time by the pinna. My prime interest is in loudspeaker playback and to avoid this problem I place the small microphones just on the side of my head, on the frame of a pair of reading glasses [6].
Pinna effects are thus avoided. The shadowing effect of the head is preserved for high frequencies and for producing a natural delay and phase shift between the ear signals at lower frequencies and for different angles of sound incidence. Loudspeaker playback now sounds natural in timbre and spaciousness and has served me as a reference during loudspeaker design and evaluation.

Playback over headphones, though, suffers from unacceptable amounts of spatial distortion. The auditory scene exists primarily between the ears and is difficult to localize outside of the head, especially in the all important frontal direction. Even for the best of binaural recordings, with a person wearing microphones at their blocked ear canal entrances, and a video camera mounted on their head to provide visual cues consistent with auditory cues due to head movement, there remains still a large amount of spatial distortion. Sources are localized outside of the head except in the front where they appear inside the forehead. In general then, sources sound too close. Their visual distance is much greater than their auditory distance. That observation is dramatically changed and sources are freely localized outside the head, if the listener's ear signals change instantly and in a natural way with his head movement. To accomplish this requires automatic head tracking [7] and

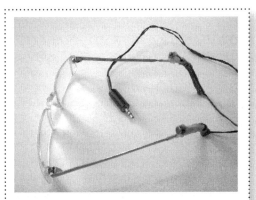

Fig. 2 – Microphone capsules mounted to reading glasses for head-related recordings

fast processing of the ear signals to provide the natural auditory cues from which we normally derive direction and distance of a sound source.

For greatest spatial fidelity in the absence of head tracking we must record the sound pressure variations at the eardrums of an individual and then reproduce at a later time the same pressure variations again at those same eardrums. This has been done successfully [5]. A small microphone, with a sound conducting flexible plastic tube in front of it, is used for recording. The tube is inserted into the ear canal where its open end is placed next to the eardrum, Fig. 3.

Playback is via supra-aural headphones, which have been equalized for a flat frequency response of the sound pressure, from recording eardrum to playback eardrum. It has been reported that this system even allows externalizing frontal sounds for some listeners for whom the system is not exactly equalized.

The eardrum binaural approach to recording and reproduction of an acoustic event is useful for scientific purposes but otherwise not practical. With loudspeaker reproduction in mind let's look at common stereo recording methods.

4 – Conventional recording and playback

My seat in Symphony Hall is in row 25, which would be too distant from the orchestra for producing a commercial recording. At this location the direct sound level is not high enough compared to the reverberant sound in the hall. I am always amazed, though, how the ear-brain hearing apparatus handles such situation much better than a microphone-loudspeaker-room reproduction system.

I hear great clarity and closeness of the musical instruments. Their reverberation conveys a sense of space but is otherwise not noticed in any unpleasant way. Yet on loudspeaker playback there is too much reverberation and a loss of clarity. Recordings are normally done from much closer to the orchestra and today mostly by using many microphones close to the musicians or their groupings.

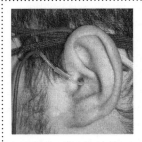
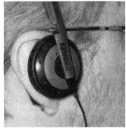

Fig. 3 – Recording and playback with identical sound signals at the eardrums

Fig. 4 – Microphone setup for an audience perspective of the orchestra

Let's look at a simple two-microphone recording from the center of row 5 and how those microphone signals, when reproduced over two loudspeakers in a living room, provide an illusion of space, distance, size and timbre of the original live orchestra. Let's assume that each microphone capsule has the directional characteristics of a cardioid, that the capsules are separated by 17 cm, and that they subtend an angle of 110°. Thus the microphones are setup in an ORTF configuration [8], so named after the French Broadcasting Authority that developed the technique, Fig. 4.

For playback let's assume a standard setup of two loudspeakers in a room. The speakers are at some distance from the walls. Loudspeakers and listener form an equilateral triangle. The loudspeakers are located at +/-30° in front and at around 3 m distance from the listener, Fig.5.

The recorded left and right microphone signals have been stored, extracted from the storage medium, amplified and fed to left and right loudspeakers respectively. What do we hear now? How similar is that to what we might have heard at the microphone position in Symphony Hall? To answer those questions I must introduce the Stereo Recording Angle (SRA) [9]. The SRA describes how sounds, arriving from any horizontal angle at the microphones, are perceptually mapped to the 60° range of angles between the loudspeakers and into left and right loudspeakers. The microphones register,

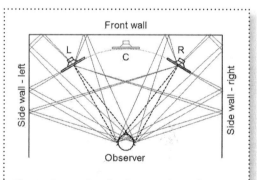

Fig. 5 – Conventional 60° – triangle configuration of stereo loudspeakers and listener in a room

of course, also sound waves from vertical angles and these too contribute to what is being heard from the loudspeakers.

On playback sound waves only emanate from the loudspeakers. Depending upon the spatial cues that are imbedded in these waves, which the brain receives from each ear, we readily imagine that some of the sounds arrive from directions other than the loudspeakers, from between or even from outside the loudspeakers. The actual loudspeakers may not seem to be the source of those sounds even when the acoustic image appears in their direction.

4.1 The perception of real and phantom sound sources

Human hearing is highly developed to identify the direction from which a sound is coming, its distance and meaning. It is an essential survival mechanism that must work in the presence of many sound sources. All these sound sources have their individual reflections and reverberation, which are characteristic of the physical environment in which the sounds occur. Our hearing sense must have worked well, whether in a cave, a forest or out in the open grasslands. We probably would not be here otherwise. Much of the initial processing of a new sound is done in the first 25 ms of its occurrence to determine direction and distance, followed by attention to the stream of sound from the new found source in order to decide whether to ignore it, hide, flee or fight. All this must happen reliably in the presence of background sound streams that are familiar and safe. The same automatic hearing mechanisms are at work when we listen to two loudspeakers in a room with reflections and reverberation and when we try to hear in a recording the direction of sound sources, their distances, reverberation and spatial context [10].

The electrical output signals from the two cardioid microphones in the ORTF setup differ in amplitude and phase from each other depending upon the angle of incidence of the sound waves. The amplitude difference is due to each microphone's directional radiation pattern pointing outwardly at +/-55⁰ from each other. The phase or timing difference is due to the 17 cm separation between the capsules, which is about the spacing between the ears. This embeds familiar timing cues into the 2-channel recording.

Let's take the setup of Fig. 5 and first apply identical signals to L and R loudspeakers. This presents an auditory dilemma to the listener because there is no precedence for such an event in the evolution of natural hearing. There have never been two identical sound sources in symmetrical directions and at equal distances from a listener. The only plausible explanation is the existence of a source in front of the listener. But at what distance? Here the room reflections give us a cue about where in space the phantom source must be located, namely at the listener's distance from the loudspeakers. In an anechoic chamber he would have localized the center phantom source as being inside his head. The listener's dilemma is readily resolved when he moves a small amount to the right or left and the phantom source quickly collapses into the right or left loudspeaker [11, 12]. The general relationship for phantom source placement between the loudspeakers is given in Fig.6. The curves are somewhat dependent upon signal type such as clicks, speech or music.

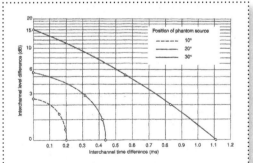

Fig. 6 – Angular position of phantom sources between two loudspeakers depending upon amplitude and time differences between the microphone output signals

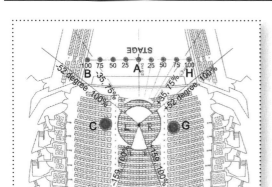

Fig. 7 – Sound pickup from ORTF microphones in a concert hall

We can see from the graph that if the sound output from L is 15 dB higher than from R and the two signals have 0 ms time difference, then the sound will appear to come from -30^0, i.e. from the left loudspeaker. If L drops to 5.5 dB it will appear to come from -20^0, and for 2.5 dB difference it will have moved to -10^0 off the center axis. If L and R are of equal amplitude but L leads by 1.1 ms, then again the sound appears to come from loudspeaker L. If the time lead is only 0.2 ms, then the phantom source is at 10^0 to the left of center. The curves show how a monaural signal can be panned to any position between the loudspeakers by amplitude or time differences between L and R drive signals. The curves are basic for the mix-down and assembly of multiple mono sound tracks and separate microphone outputs into a 2-channel stereo recording as is commonly done in popular music and classical recording.

4.2 From concert hall to listening room

Now I want to show how the two ORTF microphone outputs are mapped to the line between the loudspeakers.

A look at the floor plan of the concert hall with the microphone location in row 5 can give us some idea of the nature of the sounds coming from different directions and being picked up by the microphones, Fig.7. Upon reproduction all those sounds will appear to come either from left or right loudspeakers or as phantom sources from within the 60^0 angle between the loudspeakers in the listening room. Several investigators have worked out the mathematical details and I will only show the results [13, 14].

The ORTF microphone setup has a 100% Stereo Recording Angle (SRA) of $+/-52^0$, which is just slightly less than the 110^0 subtended angle between the microphones. 100% SRA means that all sound sources within this angle are perceptually mapped to positions along a horizontal line starting at L and ending at R, Fig. 8. Likewise 75% SRA means that sound sources within this angle are mapped as phantom sources along $+/-75%$ of the line between the loudspeakers.

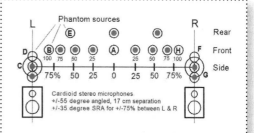

Fig. 8 - Perceptual mapping of left and right ORTF microphone signals to the loudspeakers

The microphones also pick up sounds from the sides, from the rear and from above. On playback all sources appear to originate from within the 60^0 arc in front of the listener. Specifically, sounds from the region between 52^0 and 159^0 to the left of the microphones will appear to come from the L loudspeaker. Correspondingly signals from between -52^0 and -159^0 will appear to come from the R loudspeaker. The output amplitudes from the two microphones differ so much for these angles that sound is only heard from the loudspeaker that receives the larger signal. Finally the region and signals between $+159^0$ and -159^0 behind the microphones will produce phantom sources between L and R, thus showing up in front of the listener.

During a recording session there preferably exist no specific sound sources at angles greater than $+/-52^0$ and any pickup from this region consists of reverberation only. The major portion of the re-verberated sound appears as mono signal in L and R loudspeakers. The two loudspeaker outputs are decorrelated, similar to the outputs from two widely spaced omni-directional microphones, and produce a highly diffuse image. A narrow range of angles behind the microphones generates stereo pickup of reverberated and discrete sounds appearing between the loudspeakers. The sound sources in front of the microphones are distributed between the loudspeakers in approximately correct proportions, but the sources to the right and left, to the rear and from above the microphones are not mapped spatially correct for loudspeaker reproduction.

A listener at the microphone position in the auditorium processes the signals arriving from various directions differently and can easily focus his attention upon the direct sound from the orchestra; everything else is background. A listener in front of the loudspeakers hears a spatially distorted reproduction where direct and reverberated sounds all arrive from the front. Unable to spatially dif-ferentiate between direct and reverberated sound the perceived amount of reverberation is higher during loudspeaker reproduction than it was during the recording. In addition the sound streams from the two loudspeakers are reflected by the surfaces and objects in the listening room add-ing yet another layer of sound that the mind needs to process.

The ORTF recording technique can capture the orchestra as a whole and in proper spatial and ensemble relationship of the musical instru-ments to each other. To be successful the re-cording must use a venue with reverberation acoustics that will not overwhelm the direct sound upon loudspeaker playback, lose clar-ity and sound too distant. Clarity appears to be

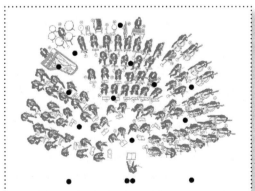

Fig. 9 – Using multiple microphones to record an orchestra

the prime objective of most recording engineers today. Spaciousness and spatial relationships are much less important to them, probably because the typical monitoring situation performs rather poorly in this respect.

The highest degree of clarity is obtained by placing microphones close to the performers or their instruments, Fig.9, [4]. The output voltages from many microphones are then mixed down to two tracks for stereo using the dependencies of Fig. 6. Spaciousness is added to the mix by artificial reverberation, but the spatial relationship between instruments often sounds unnatural. The sound of the final product is determined by the producer's expectations, the interpretive choices of the recording engineer and the performance of the musicians. Recording has become its own art form and is not merely about reproducing what a person might have been able to hear at the recording venue. Listeners who are familiar with live music, with unamplified sounds, like voice, choir, jazz, symphony, brass, etc., readily recognize whether it sounds natural when they hear a loudspeaker reproduction. For full enjoyment the microphones must have either captured a believable spatial perspective or it must have been artfully reconstructed in the mixing process.

5 – The Listening Room

Much has been written about the listening room and the placement of loudspeakers [2]. Absorbers and diffusers are frequently employed to cure acoustic problems observed with loudspeakers in the room. I have not seen those components used in living rooms to improve conversation or to make the room acoustically more comfortable when there were no loudspeakers. Unless one lives with bare walls and floors there is rarely a need for acoustic treatment. The stuff of life, upholstered furniture, rugs, book shelves, picture frames, curtains, etc, that make a room livable and not like a reverberation or an anechoic chamber, also make for a suitable listening room. The acoustic problems that are ascribed to the room are more often than not related to the loudspeaker radiation pattern and the placement of the loudspeakers in the room. Even a small room can work well provided that the listening position is proportionally closer to the loudspeakers and that the loudspeakers are closer together.

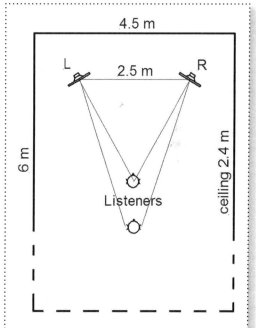

Fig.10 – Symmetrical setup of loudspeakers and listeners in a room with 1 m minimum distance from the walls

However, the loudspeakers must be acoustically small in size over most of their frequency range to exhibit a smoothly behaved radiation pattern. In the limit the loudspeakers are so close that they act like large headphones at +/-45^0 out in front, but they can still produce a large and believable auditory scene. I have two fundamental requirements for the optimum setup of stereo loudspeakers and listener in a room: symmetry and a minimum distance from reflecting surfaces, Fig.10, [15]. The listener sits equidistant from each loudspeaker, optimally at the apex of an equilateral triangle, so that the phantom sources are spread evenly between L and R as intended in the recording. Sound reflections from the room surfaces become more significant at greater distances and reduce the precision of phantom imaging. The loudspeakers themselves should be placed symmetrical with respect to large reflecting surfaces and at least 1 m away from them. This ensures that the images do not shift to the side with the nearest reflections. The delay of 6 ms or more for the reflections, compared to the direct sound at the listening position, appears to be essential for the ear-brain processor to segregate the different sound streams and to withdraw attention from the listening room. This in turn makes the recorded venue space more audible, especially when the room reflections have a neutral spectrum, which means that they are delayed and lower amplitude copies of the direct sound streams.

Appropriate placement of the loudspeakers out in the room and not in corners or hidden behind furniture may cause domestic conflicts over visual versus acoustic values or traffic flow practicalities. Small loudspeakers that are light and easily repositioned for optimum listening may be the solution though they tend to have lower volume capability and low frequency reach in the bass. Still, some designs provide a sonic balance that is satisfying and enjoyable even in those situations [18].

6 – The Loudspeakers

The loudspeakers, the source of the sound in the room, carry the major responsibility for what we hear when a recording is played. The room contributes significantly, but if the loudspeakers properly stimulate it, then our perceptual apparatus can produce a believable illusion that is primarily derived from the direct sound of the loudspeakers. Loudspeaker design relies upon science and engineering and is guided by comparing the results to live acoustic sounds. A loudspeaker must translate streams of voltage fluctuations at its input terminals into directly proportional air pressure variations for at least one point in 3-dimensional space. It should do so with minimal addition to or subtraction from the input signal. I have found that attention paid in loudspeaker design and construction to the following five areas can lead to highly believable stereo reproduction in domestic size rooms:

1.   Frequency response and stored energy
2.   Controlled directivity and diffraction
3.   Adequate volume displacement
4.   Low non-linear distortion
5.   Active crossovers and equalization

The five areas are highly interactive. Controlled directivity usually means small driver size, insufficient volume displacement, and high non-linear distortion, or large numbers of drivers. Adequate volume displacement means large drivers with limited frequency range, stored energy, and increased directivity, but low non-linear distortion. The engineering task then becomes to find the best-suited drivers, how to combine them on a properly shaped baffle and how to drive them electrically. All this with attention to the significance of any tradeoffs in the following five critical areas.

6.1 Frequency response and stored energy

On-axis frequency response is by far the most important parameter of a loudspeaker. It should be flat, whether for a recording monitor at the mixing console or for a loudspeaker that is placed in the living room. It should be flat under anechoic conditions. This will ensure some degree of sonic consistency when listening to the same recording in different locations. It is a necessary but not sufficient condition, because the off-axis radiation from a loudspeaker contributes to what is heard in a reverberant environment, as most rooms are. A flat on-axis frequency response cannot tell the whole story. It may hide low amplitude resonances that become audible only because they stretch the duration of specific tones. The mechanical structure of a loudspeaker cabinet with its distributed masses and compliances presents ample opportunity to resonate and thus to store energy at specific frequencies, which is then released gradually.

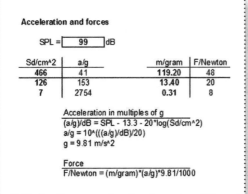

Acceleration and forces

SPL = 99 dB

Sd/cm^2	a/g	m/gram	F/Newton
466	41	119.20	48
126	153	13.40	20
7	2754	0.31	8

Acceleration in multiples of g
(a/g)/dB = SPL - 13.3 - 20*log(Sd/cm^2)
a/g = 10^(((a/g)/dB)/20)
g = 9.81 m/s^2

Force
F/Newton = (m/gram)*(a/g)*9.81/1000

Table 1 – Piston acceleration for a given SPL and piston area Sd. Motor force required to accelerate the effective piston mass.

The moving parts of a loudspeaker have relatively high mass compared to the very low mass of the volume of air particles that must be accelerated to generate a desired sound pressure level (SPL) [16, 17]. It is a very inefficient process requiring large amounts of electrical energy, about 95% of which is converted into mechanical energy and heat. For example, to obtain 99 dB SPL from the 466 cm² cone area of a 12" woofer driver requires an acceleration of 41g, or 41-times earth gravity, Table 1. If all of the woofer's moving parts have a combined mass of 119 gram, then the motor has to develop a force of 48 Newton, which is equivalent to the weight of a 5 kg mass. Sizeable accelerations and forces are required, especially for a tweeter with its light and small moving structure. The same 99 dB SPL would require 8 N of motor force to accelerate its dome and voice coil to an amazing 2754 g.

Large amounts of structure born mechanical vibration energy must be dissipated without making sound, which means without moving surface areas and accelerating air particles. Drivers are tightly

mounted to cabinets or open baffles and vibration energy travels from the drivers into those structures where it tends to excite panel resonance modes. Panels may radiate sound very efficiently at some mode frequencies, especially when the effective area is large compared to Sd, thus requiring only small amplitudes of displacement. Resonant modes store energy and release it gradually, thus lengthening the duration of those spectral components in a sound, which coincide with the mode resonance frequency, Fig. 11.

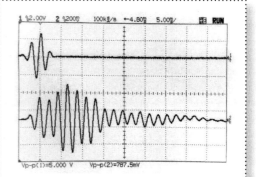

Fig. 11 – Example of energy storage in a speaker cabinet. Side panel acceleration as function of a 440 Hz burst applied to the loudspeaker.

For example, I measured the acceleration in the center of the side panel from a small 2-way box loudspeaker of 20 cm width, 29 cm height and 23 cm depth. The two sides have a total area of 1334 cm². The cone area Sd of the woofer is 104 cm² or 8% of the two sides. With 50 Vpp applied to the speaker terminals the SPL at 1 m becomes about 104 dB and the cone acceleration 330g. For the side panels I measured an acceleration of 1.02g using an accelerometer with 8.7 mV/g sensitivity and a 30 dB preamplifier, Fig. 11. Thus the radiation from one side panel would be 1.02/330 times or 50 dB below that of the cone, if the effective radiating area has the same size as Sd. The two side panels are likely to flex in and out in-phase, bringing the enclosure contribution to -44 dB. It is also likely that more than 8% of the box surface contributes to delayed re-radiation. The box panel contribution could be 30 dB below the on-axis output and would still not be noticeable in an on-axis frequency response measurement. In this example it was clearly audible as change of the toneburst sound. Knock on your loudspeaker box and you can hear the timbre of possible re-radiation.

To combat delayed re-radiation cabinets must be braced internally, which increases panel stiffness, shifts mode frequencies higher and reduces displacements. Similarly an increase in panel thickness and mass changes mode frequencies and reduces displacements. My PLUTO loudspeaker [18] avoids the panel problems by using a plastic pipe for the enclosure, which is structurally very stiff and damped. The pipe, though, has internal air resonances. Fortunately they can be sufficiently attenuated by the length of absorbing material in the pipe, through which the rear wave must travel before its remainder will exit through the thin driver membrane. Open baffles [22] also store energy, but re-radiation is less efficient because of dipole cancellation.

As much sound energy is radiated into the air cavity behind the driver cone as into the space in front of the cone. The driver cone itself is only a thin sound barrier and transmits what has not been absorbed behind it. Stuffing the box must attenuate internal air borne resonances. Furthermore the extremely high sound pressure levels inside the cabinet tend to excite the walls at certain modal

resonance frequencies. Open-baffle loudspeakers have far less serious issues with air borne and structure borne vibrations, which are reasons why they do not sound like box loudspeakers.

Energy storage is widely used in loudspeaker design to extend the response to lower frequencies and it is difficult today to find a box loudspeaker without a vent. The air in the vent behaves like a mass and the air inside the cabinet acts like a spring. Together they form a spring-mass resonator that is coupled to the woofer cone, stores energy and radiates it via the vent to the outside world. Properly tuned this system provides deeper bass than the same driver in a sealed box of similar size. It may also reduce non-linear distortion. The steeper low frequency roll-off rate comes with a peak in group delay and energy storage, which affects the naturalness of deep bass sounds. Unfortunately the common use of vented loudspeakers has generated an expectation of what reproduced bass is supposed to sound like. Many audiophiles have lost familiarity with unamplified acoustic bass.

A woofer that is tuned to a roll-off rate of 12 dB/octave, with Q=0.5 for its second order highpass response, exhibits an optimum transient and roll-off behavior. A sealed box woofer can be read-ily designed for such response. For a given driver the closed box response may roll off at too high a frequency. This can be remedied by equalization with a so-called Linkwitz Transform (LT) [19]. Ultimately the desired SPL and the woofer's maximum excursion capability determine the extent to which the cut-off frequency can be lowered. A dipole woofer would normally roll-off at 18 dB/octave, but equalization with the LT can forced it to follow a Q=0.5 and 12 dB/octave highpass filter response down to very low frequencies [22].

6.2 Controlled directivity and diffraction

The sound pressure waves from the loudspeaker illuminate the listening room. The waves are reflect-ed by large surfaces and objects, are scattered, set up a reverberant sound field and low frequency room resonance modes. The degree to which all this occurs depends upon the radiation pattern of the loudspeakers, their placement in the room and the reflective, absorptive, or diffusive properties of the various room surfaces and furnishing details, Fig. 12. For a typical box loudspeaker at low frequencies, the radiation pattern is like that of a bare light bulb. The loudspeaker radiates uniformly in all directions because its cabinet dimensions and woofer diameter is much smaller than the radiated wavelength. The specific shape of the loudspeaker matters very little.

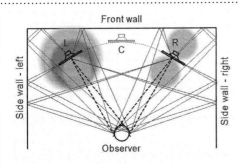

Fig. 12 – The loudspeaker illuminates the room

At higher frequencies and shorter wavelengths the edges of the loudspeaker baffle begin to diffract sound making it more forward directional. At the highest frequencies, in the tweeter range, the loudspeaker radiates like a flashlight. See loudspeaker L in Fig.12. All this is a function of the radiator and cabinet size compared to the radiated wavelength making the loudspeaker acoustically larger with increasing frequency. A radiator is acoustically small if its linear dimension x is less than 1/4th of a wavelength at frequency f and above, x < 80/f [m, Hz].

Above 4 kHz with x < 20 mm, the typical 1" dome tweeter begins to beam because it is no longer small. A 12" woofer may become directional above 300 Hz due to its own size or at even lower frequency due to cabinet edge diffraction. The loudspeaker's overall radiation pattern will change from omni-directional at low frequencies to forward radiating, often with an undesirable notch in the pattern as the acoustic output transitions from one driver to the next.

An acoustically small and open-baffle loudspeaker may have a radiation pattern that resembles a figure-of-eight at all frequencies as for loudspeaker R in Fig. 12. The room is more sparsely illuminated but with nearly the same acoustic spectrum for all directions. If box and open-baffle loudspeakers have the same on-axis frequency response, they will sound identical in an anechoic chamber but different in a reverberant environment.

It has been found by several investigators [2] that the smoothness of the frequency response curves is most important and that they should not exhibit valleys and peaks. Off-axis response irregularities can become audible via the room talking back. If frequency response curves droop in amplitude for off-axis angles, then they should decrease monotonically with increasing angle and frequency avoiding multiple beams of radiation.

The radiation pattern can be presented in different ways to show how the output sound amplitude of the loudspeaker changes with off-axis angle and frequency, Fig. 13-15.

Monitor loudspeakers as in Fig.15 are usually listened to from close distance so the direct sound dominates and the room reverberation is excluded as it is with headphones. A large

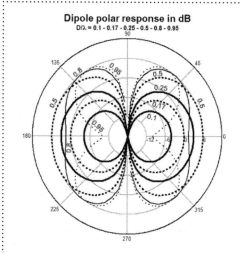

Fig. 13 – Polar response of a theoretical dipole where (+) and (-) sources are separated by a distance D. With increasing D/λ, or frequency, the output amplitude increases to a maximum at D/λ = 0.5 and the pattern widens. The on-axis response then decreases towards a minimum at D = λ.

mixing console in front of the speakers and other nearby objects can add reflections and it is important that these have similar spectral content as the direct sound. Therefore much attention has been paid to the on-axis and smoothly decaying off-axis response in the design of this loudspeaker. A recording engineer must rely on his stereo monitors to achieve an acceptable mix-down to two tracks. Hopefully the auditory scene that he constructs is spatially plausible.

While recording monitor loudspeakers are almost standardized as to their polar response, the same attention is clearly not paid to loudspeakers intended for playback of a recording in the home. For the home application, where we listen at greater distances and therefore hear more of the room, even greater attention

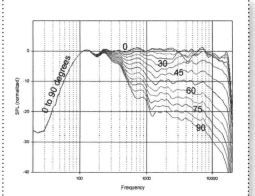

Fig. 14 – Frequency response curves for a 2-way loudspeaker with 15" woofer and 15" waveguide tweeter. The curves indicate a rapid transition from omni-directional behavior at low frequencies to near constant beam width at higher frequencies.

should have been paid to control a loudspeaker's directivity. Typically, the larger the loudspeaker becomes in physical size, the more erratic is its polar response above 200 Hz. Below that frequency all box-type loudspeakers become omni-directional and radiate more sound power into the room than at higher frequencies. This causes a spectral imbalance of the reverberant sound field in the room with emphasis upon the lower frequencies. The radiated power decreases from bass to treble by more than 10 dB. Even the best box loudspeakers exhibit this behavior, except their power response falls off smoothly with increasing frequency, without notches or peaks. Below 200 Hz is typically the frequency range where acoustic energy is stored in discrete room resonance modes, adding to the spectral imbalance. The only exceptions to the decrease in output power with increasing frequency are loudspeakers that remain acoustically small over their whole frequency range. They may have dipole, cardioid or omni-directional radia-

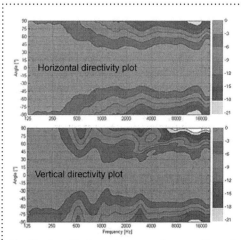

Fig. 15 – Amplitude contour map for a small monitor loudspeaker showing smooth transition from omni to wideband forward radiation both horizontally and vertically.

tion patterns and exhibit constant directivity.

It is still an unanswered question whether there is an optimum radiation pattern [20] for loudspeakers in a normally live domestic room. It is known that the pattern should be wide for perceived smoothness and naturalness of high frequencies [21]. It is also known that it should only change gradually from low to high frequencies [2]. I contend, based on my experiments and observations, that the radiation pattern should be constant and frequency independent. A loudspeaker should approach constant directivity [18]. In that case any room reflections are spectrally neutral. An omnidirectional loudspeaker fulfills this requirement, but it also illuminates the room to a maximum. A dipole loudspeaker is moderately directional and radiates 5 dB less total power into the room for the same on-axis sound pressure level as the omni-directional loudspeaker. An acoustically small dipole is my first choice, because it triggers a weaker response from the room and can be oriented to eliminate sidewall reflections [22].

A loudspeaker that is highly directional over most of its frequency range will generate fewer room reflections and can evoke an auditory scene with great clarity. The scene, though, lacks depth and openness. A center stage soloist may even be perceived as being in front of the line connecting the loudspeakers. The mind has difficulty deciphering the minimum phantom source distance from left and right loudspeaker sound streams when those have few room reflections. High directivity loudspeakers approach headphone reproduction, except that the auditory scene is heard like through a window. The window is as wide as the separation between the loudspeakers. It has little height and floats above the floor. Listening to such loudspeakers may be exciting at first but becomes tiring very soon, which indicates that the mind is hard at work subconsciously to fit the ear signals to natural patterns of sound. In contrast to that a loudspeaker with wide dispersion and constant directivity, opens up an auditory scene of similar width, but of greater height and depth. The scene is perceived as natural and plausible if the recording contains the appropriate cues. Such perception triggers a desire to hear more recordings and is not tiring. In addition to that, the spatial realism is enhanced when the loudspeakers are capable of output near live volume levels. Volume increases the size and immediacy of the auditory scene.

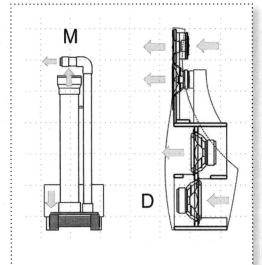

Fig.16 – Acoustically small omni and dipole loudspeakers

6.3 Adequate volume displacement

Live music can generate very high sound pressure levels (SPL) without sounding shrill, overwhelming or unpleasant. Loudspeakers often become unbearable when the volume is turned up high and that is usually the result of severe distortion. The vibrating surfaces will not move any longer in direct proportion to the electrical drive signals.

When we hear loud live music our ears will also distort, but that is a natural phenomenon and is not perceived as unpleasant. A loudspeaker should be capable of generating similar SPL as live music in order to sound realistic, but it must not contribute audible distortion of its own.

The sound pressure level that a loudspeaker generates is directly related to the volume of air that it accelerates. For frequencies where the linear size of a vibrating surface Sd is less than 1/4th of a wavelength we can assume omni-directional radiation, which allows for easy calculation of the peak displacement Xp that is required to obtain a specified SPL at 1 m from the radiator [17]. Table 2 gives the results for a 12" woofer in a sealed box (monopole) and in an open baffle (dipole).

The open baffle is of such size and shape that the rear wave from the woofer must travel a distance D before it meets the front wave. The extra distance causes a delay and thus a phase shift that is small, if D is small compared to the radiated wavelength. In that case the rear wave nearly cancels the front wave and the woofer cone must be capable of very large excursions to maintain a given SPL. The amount of phase shift increases as frequency goes up and eventually reaches 180 degrees at Fpeak, Fig. 13. Front and rear radiation add on-axis and the combined output SPL is 6 dB higher than for the sealed box at the same peak excursion. At frequency Fequal both box and open baffle have the same on-axis SPL. Above Fpeak begins the frequency region where the rear radiation subtracts again, then adds, then subtracts as the phase shift increases to 360°, 540°, 720°, etc. The radiation pattern has multiple lobes and has changed far from the low frequency figure-of-eight [23]. The practical case looks different because the dipole is not formed by point sources as assumed in the calculations for Table 2 and for Fig.12. Pure di-

Xpeak for monopole and dipole

12" woofer			Open baffle		
Sd =	466	cm^2	D =	500	mm
SPL =	99	dB	Fequal =	114	Hz
Xp =	?	mm	Fpeak =	340	Hz

f/Hz	Monopole Xpm/mm	Dipole Xpd/mm	Dip/Mon Xp ratio
20	36.3	205.9	5.7
25	22.8	103.0	4.5
32	14.4	51.5	3.6
40	9.1	25.7	2.8
50	5.7	12.9	2.3
63	3.6	6.4	1.8
80	2.3	3.2	1.4
101	1.4	1.6	1.1
127	0.9	0.8	0.9
160	0.6	0.4	0.7
202	0.4	0.2	0.6
254	0.2	0.1	0.4
320	0.1	0.1	0.4

$Xp/mm = 10^{((Xp/dBmm)/20)}$

Monopole at 1 m, free-space, dBmm
$Xpm = 37.6 + SPL/dB - 20*log(Sd/cm^2) - 40*log(f/Hz)$

Dipole at 1 m, free-space, f < Fpeak, dBmm
$Xpd = Xpm + 20*log(Fequal/f)$
$Fequal = 167*340/(D/mm)$
$Fpeak = 500*340/(D/mm)$

Table 2. – Peak excursions for a 12' woofer in a sealed box and in an open baffle for a given SPL

pole action usually ceases below Fpeak due to radiator and baffle size, as these become directional themselves. Interaction between front and rear waves disappears, but the radiation pattern remains dipolar due to front and rear beams of sound. The values in Table 2 are calculated for a 99 dB SPL in free-space at 1 m distance from a 12 inch woofer, which has a 466 cm² effective radiating area. A listening room does not represent free-space conditions and it is safe to add 6 dB to the SPL for half-space radiation due to the woofer sitting on the floor. Thus the implied target SPL equals 105 dB. The open baffle might be a flat circular panel with a radius of 500 mm and the woofer at its center, but in practice a folded baffle with the same effective D is more likely to be used in a domestic environment. The required peak excursion at 50 Hz is 5.7 mm for the sealed box, but 12.9 mm for the open baffle. It is difficult to find a woofer driver with 26 mm peak-to-peak excursion at low distortion and low turbulent air noise, as would be needed for the open baffle configuration. The solution is to use two drivers. The situation becomes worse at 25 Hz where it would take two drivers in sealed boxes and at least eight in open baffles. If this much SPL is really needed at 25 Hz, then it is more economical in size and cost to use monopole woofers below 50 Hz.

A large amount of electrical power may be required to reproduce a 25 Hz signal at 99 dB SPL from a sealed box. The stiffness of the enclosed air must be overcome to accelerate the cone to reach the required excursion Xp. The open-baffle woofer uses much less power even for its much greater excursion requirements, because it operates above the stiffness region. Here the driver's moving mass and air load determine the power required to reach Xp. Since an open baffle woofer requires more drivers than a sealed box there may not be an overall power advantage to it. I have observed and so have others that dipole woofers reproduce very deep bass more naturally in a room than sealed or vented designs [24, 25]. A convincing explanation for this observation is still outstanding, but is probably related to the modulation transfer function (MTF). It describes how much of the peak-to-trough variation of the bass waveform envelope is preserved at different measurement locations in the room. On average a dipole woofer appears to maintain greater modulation depth and is more articulate.

A 12' woofer becomes increasingly directional above a few hundred Hz. It suffers from cone break-up and its high frequency output becomes marginal. A smaller diameter driver is needed for mid frequencies and an even smaller one for the highest frequencies to maintain wide and controlled dispersion of sound. Consequently each smaller driver may have displacement issues at the low frequency end of its range. For example a 1 inch dome tweeter would have to be capable of 0.5 mm peak excursion at 1431 Hz to generate 99 dB SPL, Table 3. Two of those tweeters mounted at D = 100 mm separation from each other, pointing in opposite directions and driven as a dipole, would require a smaller Xp of 0.2 mm each. The rear tweeter's output adds to the front tweeter on-axis at frequencies above Fequal = 568 Hz for the given spacing D. Frequency and polar response problems though show up potentially above 1700 Hz where the ideal dipole response would have nulls and peaks [23].

Xpeak for monopole and dipole

1" dome tweeter			Open baffle	
Sd =	7	cm^2	D = 100	mm
SPL =	99	dB	Fequal = 568	Hz
Xp =	?	mm	Fpeak = 1700	Hz

f/Hz	Monopole Xpm/mm	Dipole Xpd/mm	Dip/Mon Xp ratio
568	3.0	3.0	1.0
716	1.9	1.5	0.8
902	1.2	0.7	0.6
1136	0.7	0.4	0.5
1431	0.5	0.2	0.4
1803	0.3	0.1	0.3
2272	0.2	0.0	0.2
2863	0.1	0.0	0.2

Table 3 - Peak excursions for a 1" dome tweeter alone and for two units in dipole configuration

6.4 Nonlinear distortion

We speak of nonlinear distortion whenever the sound pressure amplitude from a loudspeaker is not in linear proportion to the amplitude of the applied electrical drive signal. Nonlinearity in a driver causes the generation of spectral components that were not present in the applied electrical signal. The components are usually measured as harmonic and intermodulation distortion, but the measurement numbers do not fully describe the audibility of distortion. Of significance is the spectral distribution and spectral density of the distortion products. Nonlinear distortion and stored energy are first heard as a lack of clarity. At higher levels they may add harshness and unpleasantness. Only at very high levels are they perceived as distortion itself. Even amplitude clipping can be inaudible if it is of very short duration or generates few high frequency spectral components. Yet zero-crossing distortion of low amplitude signals, as found in many early solid-state power amplifiers and digital equipment, can be quite tiring. A specification of 0.1% distortion-plus-noise at 2.83 V output is not unreasonable to demand from a power amplifier. Loudspeaker drivers tend to have relatively high levels of 2nd and 3rd order distortion. The higher order distortion products are usually much lower. Distortion is generated by nonlinear changes in suspension stiffness, motor force strength and voice coil inductance with displacement Xp. The inductance also changes with drive current. This effect dominates distortion at higher frequencies. Modal breakup of the cone is yet another distortion mechanism. Driver behavior is well understood and can be measured in detail and modeled [26]. Driver design is a difficult trade-off between requirements of size, performance, reliability, repeatability and cost.

An illustration of the different forms of nonlinear distortion is given in Fig. 17. The electrical test signal is a 100 ms duration burst of a 1500 Hz sinewave that is 100% amplitude modulated at 150 Hz. The applied signal spectrum contains three frequencies, the 1500 Hz carrier and sidebands at 1350 Hz and 1650 Hz of 6 dB lower amplitude than the carrier. The acoustic spectrum of the driver's response contains many more frequencies, both above and below the excitation spectrum. Looked at in the time domain we see that the envelope amplitude of the burst decreases over its short 100 ms duration. This indicates thermal compression due to voice coil heating, increased voice coil resistance and thus reduced current flow. Additionally the driver exhibits stored energy because the 150 Hz envelope of the burst does not go to zero between the envelope maxima. The 10 W of applied power at 1500 Hz are too much to handle for this driver but it might still be useful at higher frequencies where less displacement Xp is needed for the same SPL.

Woofer, midrange and tweeter drivers must be extensively tested for their nonlinear behavior and

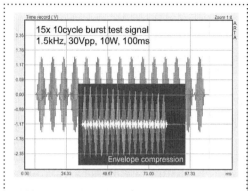

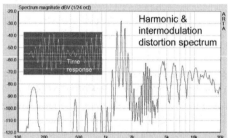

Fig. 17 – Example measurement of a tweeter using a 10 W burst of 100 ms duration to show gain compression, stored energy, harmonic and intermodulation distortion

for stored energy to determine over what frequency range they can be used at the targeted maximum SPL. Deviations from a linear frequency response can be equalized. Distortion is much more difficult to correct electronically.

6.5 Active crossovers and equalization

A properly designed 2-way loudspeaker can do a respectable job of reproducing the full audio frequency range at near realistic volume levels and with wide dispersion. Only bass extension and volume must be sacrificed. Adding two subwoofers can solve the problem, if the subwoofers can be integrated optimally with the 2-way loudspeakers to form a coherent 3-way system. In all cases crossovers are involved to divide the frequency range into bands that can be handled by tweeter, midrange/woofer and subwoofer drivers. Furthermore the signals for all driver must be equalized in amplitude and phase, to correct driver response imperfections and driver offsets, for the filter bands to add acoustically to the desired overall frequency response. Traditionally passive filter networks are inserted between the power amplifier output and the individual drivers. The purpose of the inductors, capacitors and resistors in the network is to decouple the power amplifier from a driver at frequencies, which the driver should not radiate and to equalize the frequency response over the range that the driver is used for. The power amplifier controls the mechanical movement of the driver's voice coil, but only at frequencies where it is fully connected to the voice coil and not at frequencies where the passive crossover network interferes to increasing degree. If the filter network is placed ahead of the power amplifier, then the amplifier remains connected to the voice coil at all frequencies and has maximum control. The penalty is the need for a separate power amplifier for each driver. That is a small price to pay when the power requirements for each amplifier are much smaller than for the single power amplifier that was needed beforehand. A 3-way active system with three 60 W amplifiers is not equivalent to a single 180 W amplifier but closer to a 540 W amplifier. When the 60 W woofer amplifier clips, the midrange and tweeter drivers do not see a distorted drive signal. When the 180 W power amplifier clips driving a passive crossover loudspeaker then the tweeter is the first to see it because of the high frequency components generated and you are likely hear it. Using an amplifier for each driver and an active crossover/equalizer is an effective way to reduce distortion

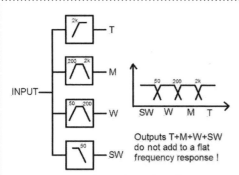

Fig. 18 – Commonly used topology for a passive crossover

Outputs T+M+W+SW do not add to a flat frequency response !

and obtain high output volume levels.

In my book passive crossovers are a waste of time, if highest performance is to be obtained from a 3-way or 4-way design. Passive crossovers are typically built as sets of parallel filters, Fig. 18. Differences in driver sensitivity can only be equalized by dissipating some of the amplifier's power in the filter networks. Also the outputs from the filters add to a flat response only if the crossover frequencies are a decade or more apart. For example, the 50 Hz highpass filter for W in Fig.18 adds phaseshift to the 200 Hz lowpass. Thus W and M do not add with the correct phase at 200 Hz. The 50 Hz and 200 Hz crossover frequencies are too close together.

An active crossover/equalizer [27] allows for easy implementation of the correct network topology, Fig. 19, [28]. Now the phaseshift from the 50 Hz highpass in the W channel is carried forward in W and M channels so that W and M add correctly and regardless of how far apart the crossover frequencies are. The dividing and equalizing filter networks ahead of the power amplifiers can be built with operational amplifiers, resistors and capacitors with precision and consistency.

The active crossover/equalizer filter functions can also be implemented with a digital signal processor (DSP). This could potentially result in superior loudspeaker performance, if the acoustic outputs from the individual drivers add in phase, as with an analogue crossover, and if the

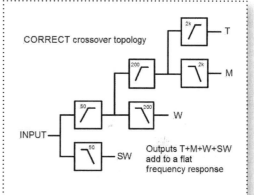

CORRECT crossover topology

Outputs T+M+W+SW add to a flat frequency response

Fig. 19 – Correct and universal topology for a crossover network

phase of the combined output changes linearly with frequency. Such loudspeaker would reproduce an electrcal squarewave signal at its input as an acoustic squarewave, though only at a single point in space, unless the individual drivers are coincident. Our hearing apparatus is quite insensitive to waveform distortion due to phase. I am not convinced that linear phase reproduction is audibly significant, when huge amounts of phase distortion have already been introduced in recording and mix. Deviation from linear phase can be audible on certain test signals but it is probably the lowest contributor to the realism of recorded sound.

8 – Other contributors to sound

Much has been written and could be said about the sound of certain power amplifiers, preamplifiers, tubes, DAC's, cables, interconnects, etc.. I have heard changes in sound from such devices. There are usually explanations for them that are dependable enough to design the effect into or out of a piece of equipment, even when it is difficult to measure what has been heard. Overall, though, the audible significance of these effects is small compared to what is contributed by microphone selection, recording technique, mix, loudspeakers and room to the spatial distortion and believability of the reproduced auditory scene. Long term wear and not initial excitement determines whether an auditory effect is accepted subconsciously as natural or as a tiring artifact. I have observed that some audiophiles listen primarily for artifacts and become involved with the music through them, if ever. This can turn into an endless search for the perfect component and an end in itself, particularly if their reference is no longer unamplified sound in spatial context.

9 - Summary

So what does it take for STEREO to sound at its best? It takes a lot of attention to detail at every stage in the signal transmission path from acoustic vibration to acoustic vibration, from recording to storage and reproduction. Sound exists in space. We naturally hear the direct and the reverberated sounds and know therefore where we are in relation to that space and the sounds in it. The best stereo system will deliver a believable illusion of hearing sound sources in their spatial context, in their natural relationship and without intrusion by the listening room and the loudspeakers. Though the presentation is in front and not surrounding the listener, it mimics much of a concert hall experience and can engage a listener to a high degree of satisfaction. This has been my observation and experience.

10 – References

[1] Linkwitz/Riley crossover filters, http://www.linkwitzlab.com/filters.htm#3

[2] Floyd E. Toole, Sound Reproduction – Loudspeakers and Rooms, Focal Press, 2008

[3] Albert S. Bregman, Auditory Scene Analysis – The Perceptual Organization of Sound, The MIT Press, 1999

[4] David Griesinger, The Importance of the Direct to Reverberant Ratio in the Perception of Distance, Localization, Clarity and Envelopment, 126th AES Convention, Munich 2009, Preprint 7724, http://www.davidgriesinger.com/

[5] David Griesinger, Frequency Response Adaptation in Binaural Hearing, 126th AES Convention, Munich 2009, Preprint 7768, http://www.davidgriesinger.com/

[6] Microphone construction, http://www.linkwitzlab.com/sys_test.htm#Mic

[7] Head-tracking, http://www.acoustics.org/press/159th/sunny.htm

[8] Schoeps ORTF stereo microphone http://www.schoeps.de/en/products/mstc64u and application http://www.schoeps.de/en/applications/showroom

[9] Michael Williams, Microphone Arrays for Stereo and Multichannel Sound Recording, Editrice Il

Rostro, 2004, http://www.posthorn.com/Micarray_1.html

[10] Jens Blauert, Spatial Hearing – The Psychophysics of Human Sound Localization, The MIT Press, 1997

[11] Francis Rumsey, Spatial Audio, Focal Press, 2005

[12] Peter Damaske, Acoustics and Hearing, Springer, 2008

[13] Helmut Wittek, Guenther Theile, The Recording angle – Based on Localization Curves, 112th AES Convention, Munich 2002, http://hauptmikrofon.de/HW/AES112_Wittek_Theile.PDF

[14] Stereo Recording Angle calculation from Image Assistant 2.1, http://www.hauptmikrofon.de/

[15] Siegfried Linkwitz & Don Barringer, Recording and Reproduction over Two Loudspeakers as Heard Live; Part 1: Hearing, Loudspeakers and Rooms, 126th AES Convention, Munich, 2009, Pre-print 7745, http://www.linkwitzlab.com/publications.htm, #26

[16] Leo L. Beranek, Acoustics, McGraw-Hill, 1954

[17] Excursion-limited SPL Nomographs, http://www.linkwitzlab.com/publications.htm, #4

[18] PLUTO-2.1 Loudspeaker, http://www.linkwitzlab.com/Pluto/Pluto-2.1.htm

[19] Linkwitz Transform, http://www.linkwitzlab.com/filters.htm#9

[20] Siegfried Linkwitz, The Challenge to Find the Optimum Radiation Pattern and Placement of Stereo Loudspeakers in a Room for the Creation of Phantom Sources and Simultaneous Masking of Real Sources, 127th AES Convention, New York, 2009, Preprint 7959, http://www.linkwitzlab.com/publications.htm, #27

[21] BeoLab 5 Loudspeaker, http://www.bang-olufsen.com/beolab5

[22] ORION+ Loudspeaker, http://www.linkwitzlab.com/orion_challenge.htm

[23] Models for a Dipole Loudspeaker Design, http://www.linkwitzlab.com/models.htm

[24] Siegfried Linkwitz, Investigation of Sound Quality Differences between Monopolar and Dipolar Woofers in Small Rooms, 105th AES Convention, San Francisco, 1998, Preprint 4786, http://www.linkwitzlab.com/publications.htm, #1

[25] Keith Holland, Philip Newell, Peter Mapp, Modulation Depth as a Measure of Loudspeaker Low Frequency Performance, Proceedings of the Institute of Acoustics, Vol.26. Pt.8.2004

[26] KLIPPEL Know-How, http://www.klippel.de/pubs/papers.asp

[27] Analog Signal Processor – ASP, http://www.linkwitzlab.com/orion_asp.htm

[28] Crossover topology mistakes, http://www.linkwitzlab.com/frontiers_5.htm#V

Project 21 Part I

Jean-Claude Gaertner

Project 21 is an extensive project consisting of a satellite two-way with a bass enclosure. The individual drivers are driven actively through a DSP-based crossover. Four-channel level controls are used at each speaker location. The levels are set remotely trough an RF link from a preamp and include volume, balance and individual channel level offset functions. This is a joint project from several people. In Part I, Jean-Claude Gaertner discusses the design considerations and prototype measurements of the satellites.

Under this somewhat enigmatic title, I present a project to build a HIFI loudspeaker system using top-of-the-range components. In view of the cost of this realization in its optimal version, I decided to structure it in such a way that, as far as possible, its elements can be used independently.

The complete project consists of the following parts:

- Two mid-high enclosures (active or passive filter);
- Two (or more) subwoofer boxes (active filtering only);
- Two digital signal processors 24 bit / 96 Ks with 2 inputs - 4 outputs;
- A radio frequency volume - balance - individual level control at DSP analog outputs.

1   P21 MID-HIGH ENCLOSURE

Until Spring 2006 my listening system was composed of an active tri-amplification system with on one hand, two bass reflex subwoofers and on the other hand, two mid-high loudspeakers based on a 6,5-inch PHL driver and a RSQ8 planar Philips tweeter, the manufacturing of which has been stopped for several years now. The sound was excellent and I liked the total transparency and the fusion between the impregnated high-strength cellulose fiber cone and the planar tweeter, but the marked directivity of the later somewhat blackened the sound, narrowing the sweet spot, especially vertically. I wondered then, whether it would be possible to find another planar or ribbon tweeter with a more acceptable polar response. After several months of internet search and reading in audio forums, only two were left in the running: the NeoPro5i from FOUNTEK[1] and the 140-15D from RAAL[2] . But as none of them were available in France I had to make a choice based on measurements that designers agreed to send me. I ended up choosing the 140-15D which offered good performances as well as a fast and easy way to change the ribbon in case of damage. In July 2006, on invitation of Mr. Aleksandar Radisavljevic, designer of this tweeter, I went to Belgrade to pick up the pair of drivers I had ordered and there I had the opportunity to have long talks about ribbon driver design as well as many other audio topics. Nearly at the same time, Mr. Philippe Lesage from PHL AUDIO[3] informed me about the improvements made to his 6,5-inch series drivers, in particular with respect to the rubber surround, flux uniformity, cone impregnation and coating.

1.1   RAAL 14015D Tweeter

Aleksandar Radisavljevic created the RAAL company in 2004. The company is based in Zajecar, Eastern Serbia, close to the Rumanian border and has expanded rapidly over the last several years. The major innovations he claims are: a homogeneous magnetic field (Equafield®), a non-corrugated thin ribbon (Flatfoil®) and a symmetrical signal routing (Symmlead®).

The working principle of a ribbon tweeter is extremely simple: a conducting ribbon is placed in a high magnetic field and coupled with a transformer so that the amplifier sees the correct impedance. For an audiophile of a certain generation, a legendary name comes immediately to mind: the DECCA KELLY tweeter which was developed by British engineer Stanley Kelly in the fifties then taken over by DECCA firm under the name DK30. Some people still speak very highly of it. However, although in listening tests the Kelly ribbon can generally be qualified as transparent, it's also associated with brittleness, low power capability and a bad polar response, which discourages many manufacturers for using it. But recently, there has been a renewed interest in this technology, at least in HiFi circles.

---

1   www.fountek.net

2   www.raalribbon.com

3   www.phlaudio.com

1.1.1 Presentation

The picture of the 140-15D gives a good outline of this ribbon tweeter which seems to be traditionally built. The dimensions of the magnet system are of 61 x 151 x 84 mm and of 90 x 180 mm for the stainless front plate (painted in black on the picture). The driver weighs 2,7 kg and the 4-micron thick aluminum ribbon measures 140 mm by 15 mm. Except a very light embossing, this ribbon is not corrugated as seen in other products. Both connections of the ribbon to the transformer are made outside the magnetic field on one side only. You can notice the vertically notched opening front plate diffusing the energy to reduce strong reflections. Replacing a ribbon is often a delicate operation, but RAAL makes available a "ribbon kit" that simplifies this operation.

The procedure consists of unsoldering the two strips, withdrawing the six screws, the front plate and the protective grill; the ribbon unit can then be taken out. It can be done very easily, having had to do it myself. As I was dismounting one driver to check its internal structure, one screw was attracted by the strong magnetic field, snatched out of my hand and stuck to the magnet through the ribbon. Therefore, be very careful when dismounting your tweeter. Don't forget that you are working in a high magnetic field that will attract metallic objects.

As I had taken the precaution to buy two spare ribbon kits it took only a few minutes to repair the damage. This unfortunate episode, which occurred at the end of the tuning process, at least made it possible to check that a ribbon replacement does not modify tweeter measurements.

What makes that tweeter unique to my knowl-

edge is the presence of two foam pads of a very particular shape on both sides of the ribbon. They are essential to its good performance and we will see their influence in the paragraph dedicated to measurements. Slightly magnetized, they hold the position you leave them in. The first specimens were delivered with the A pads (picture n°9) but, because of an important rejection rate due to variations in the alveolar structure between the batches, new pads were developed with a much more homogeneous foam structure. These pads, representing the current production, had to be reshaped (B picture n°9).

1.1.2 Preliminary measurements

I used PRAXIS software developed by Mr. Bill Waslo at Liberty Instruments Inc[4]. This measurement system along with a Digigram VXPocket or a RME Fireface 400 sound card, depending on the need, has replaced my old MLSSA DOS system which revolutionized the world of acoustic measurement in the middle of the eighties. Such a system is certainly expensive, but much more affordable alternatives, which I'll mention later, exist. They will allow you to make your own measurements and extract the maximum listening performance from any given system in your room, particularly in the low register. I also used an ACO PACIFIC[5] 7012 ½ inch microphone with a flat curve up to 40 kHz along with a 4012 preamplifier.

First of all, let's examine the influence of these two foam pads. To measure it, I mounted the tweeter in a 26 cm Width x 43 cm Height x 39 cm Depth enclosure. The microphone was positioned on the tweeter's vertical axis at a distance of 60 mm from the front plate and the distance X was made to vary from 70 mm to 40 mm.

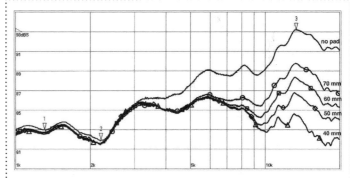

Figure 1e RAAL 140 freq response

Mkr1: 1.31kHz, 82.97dBS. Mkr2: 2.203kHz, 81.96dBS.

Mkr3: 13.3kHz, 93.2dBS.

---

4   www.libinst.com

5   www.acopacific.com

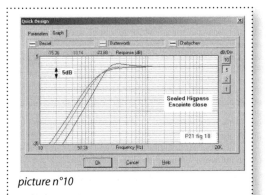

picture n°10

Figure n°1e is an instructive picture; note the vertical scale of just 2 dB per division. Speaker manufacturers very often use a 10 dB/div scale in their literature to present more flattering curves. The top curve is the 140-15D on axis response without any foam pads. Apart from a small well damped resonance of 2 dB around 13 kHz, a 7 dB increase starting at 2.5 kHz can be seen, which is quite common for a ribbon driver. The dip around 2.2 kHz does not come from the driver itself but is a diffraction effect as we will see later. These pads are able to model very effectively the response curve. From 3 kHz to 7 kHz, whatever the distance between the pads, the response is almost flat. At 10 kHz, a 10 mm variation corresponds to a 1-1.5 dB attenuation. It appears that a 50 to 60 mm distance between the pads offers the best compromise, at least on the axis, the response lying then well within ± 2 dB. The validation of this excellent result has still to be done by checking the polar response in the final enclosure. It will be very easy to adjust the level above 8-10 kHz, depending on listening room absorption and individual ear response, by simply sliding the pads towards or away from each other. Efficiency differences between the former and the current pads are shown in figure n°2e. It is noticeable that the current pads are more effective by 2 dB at 15 kHz at the equivalent distance of 50 mm, but it is striking that their curve appearance is similar to the former pads' curve appearance, so you need only to increase the distance between by 10mm to make them coincide within a fraction of dB.

Amplifiers, usually designed as voltage generators with low output impedance, can't drive low impedance sources like a ribbon speaker directly. So, the manufacturers integrate a step-up transformer that has to be well calculated not to introduce distortions or sound colorations. The impedance (figure n°3e) is presented at first approximation in the form of a resistance with an inductance in parallel. At around 7 Ω above 900 Hz, 4 Ω at 300 Hz, the driver impedance is lower than 1 Ω below 80 Hz. In the case of active filtering, it will be necessary to

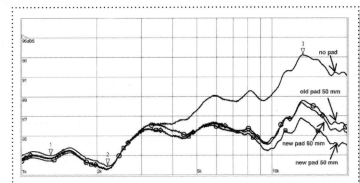

Figure 2e RAAL 140 freq response
Mkr1: 13.7kHz, 82.97dBS. Mkr2: 2.203kHz, 81.96dBS.
Mkr3: 13.3kHz, 93.2dBS.

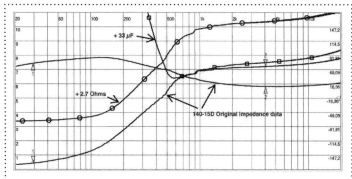

Figure 3e RAAL 140 Impedance

Mkr1: 29.3Hz, 436.5mΩ, 63.0°. Mkr2: 4.04kHz, 7.202Ω, 10.02°.

protect both the tweeter and the amplifier from overload. The simplest solution consists of inserting a low value non-inductive resistor at the expense of a reduced efficiency.

For example, with 2.7 Ω in series, the amplifier does not see a lower impedance than 3 Ω. The tweeter is protected in the low frequencies since the main part of the signal will be dissipated in the resistance. This is offset by a level reduction of about 2.5 dB. Usually, that's not a problem. I have adopted this solution for active filtering because there is neither any deterioration in frequency nor in phase response. And if this is not appropriate (low power amplifier for instance) there is another option: to insert a capacitor instead of a resistor. There is now no longer any loss in the active band but you instead have to deal with a first order high-pass electrical filter. The best compromise I have found is a 33 uF value. Its contribution can be clearly seen below 550 Hz (figure n°3e). The impedance does not go lower than 6.5 Ω, both the amplifier and the tweeter are protected, but it introduces an additional constraint for active filtering. Of course, in the case of passive filtering, the problem will disappear since we will need to introduce at least one capacitor to create a high pass filter.

Distortion (figure n°4e) is very low with less than 0.1% for the 2nd harmonic which dominates down to 1 kHz. The 3rd harmonic remains constant and at a lower level down to 2 kHz, then increases regularly to exceed the 2nd harmonic at 1 kHz and below. From the distortion point of view, it is desirable to choose a cut-off frequency in the 1.5 - 2 kHz range.

1.2 PHL AUDIO SP1280 Mid Driver

As soon as I got the RAAL tweeters, I couldn't resist testing them in place of the RSQ8 in the enclosures I was using at the time. I quickly built a small plywood plate for the new tweeters, did some measurements to

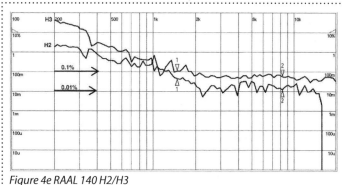

Figure 4e RAAL 140 H2/H3

Mkr1: 1.478kHz, 1.103%, 0.044%. Mkr2: 8.264kHz, 0.055%, 0.012%.

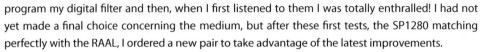

program my digital filter and then, when I first listened to them I was totally enthralled! I had not yet made a final choice concerning the medium, but after these first tests, the SP1280 matching perfectly with the RAAL, I ordered a new pair to take advantage of the latest improvements.

Mr. Philippe Lesage, former chief of AUDAX research lab, created PHL in 1990. After reorganization in 2004, PHL has become a leader in its field. Note that Professional audio and HiFi are two entirely different worlds because their constraints differ. Very often HiFi refers to trendy design, esoteric products, fancy cables, to name but a few, coupled with sometimes high prices. HiFi manufacturers try to impress with marketing innovations and distinguishable arguments without offering any measurable improvement. As Professional Audio implies involvement at an industrial level, high power handling capacity, unconditional stability in time, resistance to various climatic conditions and especially reliability are constraints impossible to circumvent. Generally speaking, a new technology is adopted only if it brings a true improvement. Innovations, when they exist, are seldom seen, but hidden in motors, surroundings, cone assembly and resins. PHL AUDIO started by developing an innovating series of 6,5 inch drivers which quickly ensured its notoriety before extending the range up to 18 inch. Good news for 2010, tweeters are announced. It's a pity that their website does not reflect anymore the extent of their range, but working primarily according to customers specifications, updating its content seems not to be their priority right now.

1.2.1 Presentation

Picture n°11 gives a good idea of this loudspeaker. The gasket, with a maximum obstruction of 187 mm, is molded in a high tensile alloy. Its form has been studied to provide a maximum rigidity while also being used as a heat spreader (intercooler® system). The 1.5-inch voice coil uses a rectangular shape copper plated aluminum wire wound on a vented Kapton®- glass polyamide support.

The exponential fiber cone is made up of high-strength cellulose impregnated and coated on both sides with damped resins to increase the Young modulus and bring the climatic resistance at a tropical level. The operational range is guaranteed from - 10 °C to + 50°C. The magnetic system, comprising a 12 cm ferrite magnet, has been optimized for field symmetry and equipped with a flux stabilization ring. A black anodized aluminum phase plug regularizes the acoustic load in the upper part of the frequency spectrum.

1.2.2 Preliminary measurements

I extracted main Thiele / Small parameters by using the additional mass method. For these measurements, and in accordance with IEC standards 60268-1, I carried out a one-hour preconditioning, followed by a one hour recovery period. I used an 11 Ω resistor and an output voltage giving a 10 mA current at 200 Hz.

The table below presents these parameters calculated directly by

Picture n°11

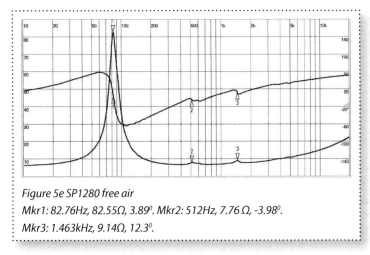

Figure 5e SP1280 free air

Mkr1: 82.76Hz, 82.55Ω, 3.89°. Mkr2: 512Hz, 7.76 Ω, -3.98°.

Mkr3: 1.463kHz, 9.14Ω, 12.3°.

PRAXIS but also with the LEAP[6] tool "Transducer Model Derivation" by importing the two impedance curves measured with PRAXIS.

*LEAP 5 uses different parameters (Krm, Erm, Kxm, Exm) to model impedance variations.

Although both programs use different extraction methods, they give almost the same T/S parameters. We can reasonably assume that these parameters are valid. Indeed, it never should be forgotten that we work with very low signal amplitudes and many errors can spoil the results. It is advisable to carry out several measurements to check the relevance of the results.

The SP1280 with a resonance frequency of 82 - 84 Hz and an efficiency of 92 - 93 dB is truly a medium range driver. Two small accidents (500 Hz and 1450 Hz) can be seen in the free air impedance curve (figure n°5e). The one that occurred at 1450 Hz is probably due to a mode change in cone operation but I've no answer concerning the one at 500 Hz.

1.2.3 Use of T/S parameters

The LEAP 5 "Quick Design" utility makes it possible to quickly get an idea of how the driver performs

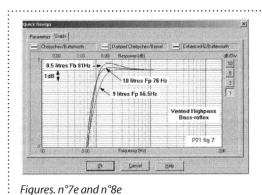

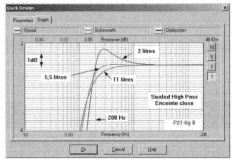

Figures. n°7e and n°8e

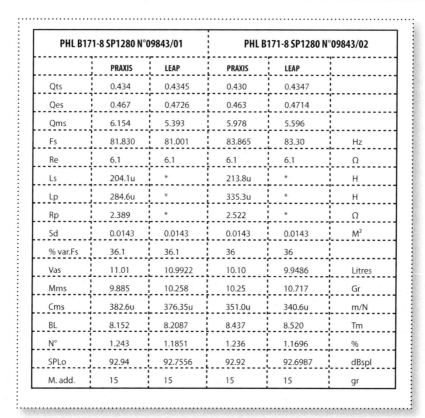

PHL B171-8 SP1280 N°09843/01			PHL B171-8 SP1280 N°09843/02		
	PRAXIS	LEAP	PRAXIS	LEAP	
Qts	0.434	0.4345	0.430	0.4347	
Qes	0.467	0.4726	0.463	0.4714	
Qms	6.154	5.393	5.978	5.596	
Fs	81.830	81.001	83.865	83.30	Hz
Re	6.1	6.1	6.1	6.1	Ω
Ls	204.1u	*	213.8u	*	H
Lp	284.6u	*	335.3u	*	H
Rp	2.389	*	2.522	*	Ω
Sd	0.0143	0.0143	0.0143	0.0143	M²
% var.Fs	36.1	36.1	36	36	
Vas	11.01	10.9922	10.10	9.9486	Litres
Mms	9.885	10.258	10.25	10.717	Gr
Cms	382.6u	376.35u	351.0u	340.6u	m/N
BL	8.152	8.2087	8.437	8.520	Tm
N°	1.243	1.1851	1.236	1.1696	%
SPLo	92.94	92.7556	92.92	92.6987	dBspl
M. add.	15	15	15	15	gr

under three different alignments.

For the three vented high-pass alignment types, the volume varies only between 8.5 and 11 liters at a - 3 dB cut-off frequency between 85 to 95 Hz. For the sealed high-pass alignments, optimal volume appears to be around 5 - 6 liters for a Butterworth damping, the cut-off frequency is then at 140 Hz. So, which one would be the best to start with? At first sight, a vented enclosure (9 liters figure n°7e) seems preferable. But let's examine the same curves now, but with a 5 dB vertical scale (figures n°9e and n°10e).

It is noticeable that below 100 Hz, the sealed high-pass stop-band slope is twice as low as the vented one. In short, there's a choice to be made between a higher cut-off frequency (about 140 Hz) with a 12 dB asymptotic stop-band slope or a lower one (about 90 Hz) with a double stop-band slope.

1.2.4 Low frequencies simulation

Using the LEAP extended T/S parameters functionality, I simulated the response and excursion of the SP1280 in a 5 liter-closed box at 100% fill with 20 kg/m³ fiber glass wool and in a 9 liter-66 Hz-tuned vented box filled with 50% of the same fiberglass wool at 1 Watt (figures n°11e and n°12e).

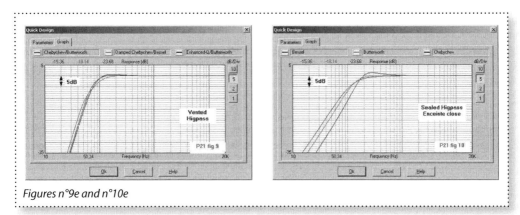

Figures n°9e and n°10e

From the sound pressure side, the vented enclosure takes the lead up to 39 Hz but gets 12 dB less pressure at 20 Hz. On the excursion side, at 100 Hz, the vented enclosure offers 3 dB more gain for the same cone displacement and preserves its advantage down to 56 Hz thanks to the vent radiation. On the other hand, at 40 Hz and at the equivalent level of 74 dB, the cone moves by 1 mm compared with the 0.57 mm of the closed box. Let's see now what occurs while increasing the power.

The vented enclosure

The figures n°13e and n°14e show the SPL and excursion versus frequency by increasing the power by 3 dB steps. The linear cone displacement of the SP1280 is ± 3.5 mm and ± 8 mm before damage. I set the reference mark at 3 mm to keep a small error margin. You can see in figure n°14e that this limit is reached at 59 Hz for 64 Watt and 39 Hz for 8 Watt. There is also a brutal increase of displacement

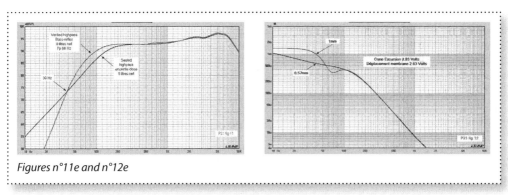

Figures n°11e and n°12e

below the tuning frequency. It is a normal behavior for a vented enclosure because the vent does not radiate anymore and the driver is no longer properly loaded by the back volume. Figure n°13e shows the 3 dB regular increase while doubling the applied power until 32 Watt, then the heating of the moving parts and of the motor modifies the transducer characteristics as you can see on the 64 Watt curve. In "A" we can see an overshoot due to these modifications; in "B" we note a compression,

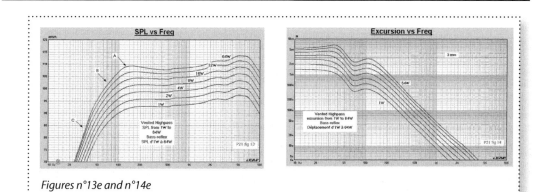

Figures n°13e and n°14e

as the air flow in the vent is not linear anymore (for this simulation I used a circular vent of 63 mm diameter which is realistic for a 9-liter box); in "C" the linear displacement is largely exceeded. These simulations show that we will need to electrically filter the high-pass vented enclosure at 60 Hz or above for the driver to stay in the linear domain. It will then be possible to reach a very comfortable level of about 110 dB at 1 meter above 150 Hz.

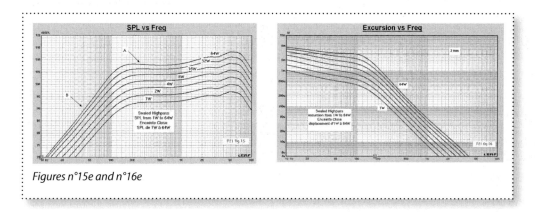

Figures n°15e and n°16e

The closed box

At 64 Watt, the displacement reaches 3 mm at 100 Hz but remains at this value down to nearly 50 Hz, before increasing slowly to reach 4 mm at 20 Hz (figure n°16e). At 32 Watt the displacement remains in the linear field down to 30 Hz. There are also some modifications on the SPL curve at 64 Watt, but at a lower degree (point A figure n°15e). In "B" the linear displacement has been exceeded too, but the deformation is less marked than in the preceding case, thanks to the back volume load which helps control the driver with increasing power. An electrical crossover at 50 Hz or above will let the driver stay comfortably in the linear domain.

1.3 The Box

The choice of a box shape, its volume and type of load is always the result of a compromise. Here, preliminary measurements taken in a 10-liter enclosure showed that the tweeter with its very wide polar response was very sensitive to its close-in environment. That implies that it is necessary to flush mount it to minimize close-in diffractions and to choose a well-adapted box shape. The medium driver has a good polar response up to 2 kHz (which is logical with a 135 mm diameter exponential cone) but to keep a good homogeneity in space with the tweeter, the front face has to be reduced to the strict minimum. And in addition, aesthetic criteria shouldn't be totally neglected either. With the use of 18 mm Finland Birch plywood or 19 mm MDF panels, and considering the volume occupied by the drivers and the reinforcement frame, the dimensions of the front face cannot be smaller than 19 cm by 39 cm. To reduce the box influence on the polar response and to minimize internal reflections while not complicating the realization too much, I chose a trapezoidal shape for the closed box. For the vented enclosure, I suggest to keep the rectangular shape which makes it possible to limit the depth extension to 19 mm.

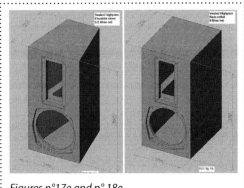

Figures n°17e and n° 18e

Figures n°17e and n°18e present two possible versions of this enclosure. Their front face dimensions are 196 mm by 390 mm. The depth is 216 mm for the sealed high-pass box version and 235 mm for the vented high-pass box, the vent being on the rear. Net volume is 5.5 liters and 9 liters respectively, after deduction of the volume occupied by the drivers, the reinforcement frame, a possible passive crossover and the vent.

1.3.1 What type of load to choose?

I had used a vented high-pass enclosure for my previous satellites (SP1280 1st version and RSQ8), thinking that it was judicious to have the maximum SPL extension in the 70 Hz - 150 Hz range. But after several months of extensive listening, I ended up by blocking the vent, adding some bricks inside to reduce the volume and modifying the filtering consequently. Although, objectively it should be possible to obtain the same sound with adequate filtering, I preferred the SP1280 sound in a sealed box. A possible explanation is that the natural second order slope facilitates the transition with the subwoofer in a semi-reverberating room.

1.3.2 The cabinet

Thus, I chose to build the sealed enclosure. Picture n°13 shows a pair of these made of 18 mm-thick Finland birch plywood at a professional workshop. Most of the HiFi manufacturers use MDF instead. Studies on vibration behavior of both of these materials did not reach a clear consensus. A few years ago, I made some measurements with an accelerometer and pseudo random noise sequences and although I noted different spectrum responses without a direct relationship to listening tests, I especially noticed how important the manufacturing quality was. In fact, professionals use plywood to reduce the weight (a good third at equivalent thickness) and Finland birch in particular for its homogeneity and shock resistance. In this particular case, given its trapezoidal shape, its small size and its horizontal reinforcement frame, the nature of the material comes second. It can be built in 19 mm MDF if the 18 mm Birch plywood can't be found. A small opening in the back is intended to receive the Speakon® plug which I have used for the connections.

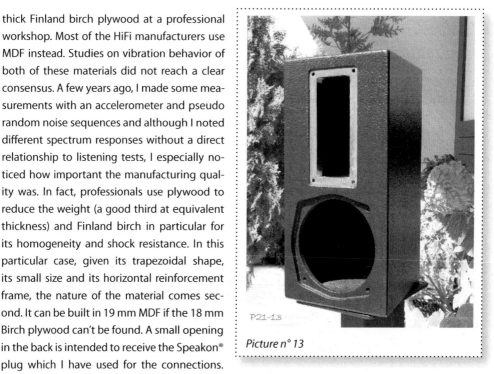

Picture n° 13

The edges have been smoothed for aesthetic reasons and not to decrease edge diffraction as it is regularly said in specialized magazines. If you look at the relationship between the wavelength λ=C/f and the curvature radius, you'll notice that to be effective, the round-off would have to be of several centimeters and would only have some influence beyond 5 or even 10 kHz. I have decided not to physically align the acoustic centers, because that would create a discontinuity in the front face, harmful to the tweeter radiation. On the other hand, there's no getting away from recessing the two drivers.

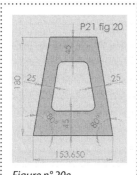

Figure n° 20e

1.3.3   Construction Plans

Figure n°19e gives the most useful dimensions to build a sealed high-pass enclosure made of 18 mm plywood or 19 mm MDF. For the recess opening of the SP1280 it is best to use the frame itself to make a gauge for the router.

The horizontal frame (figure n°20e) is placed between the tweeter and the medium driver to reinforce the enclosure. It is important to respect the 201 mm dimension (figure n°19e) as there is little room left for the inserts of the fastening screws, if you use some. In that case, it will perhaps be necessary to place the two lower tweeter inserts before posi-

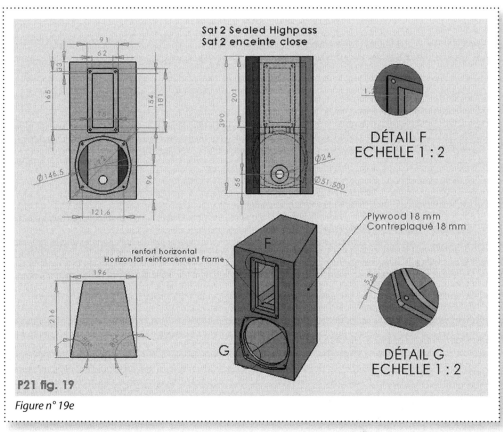

Figure n° 19e

tioning the reinforcement frame. Be careful, because the dimensions given in figure n°20e apply for 18 mm thick plywood, remember to correct them if you use 19 mm MDF. The recess for the tweeter front plate is 1.2 mm (detail F) and 5.3 mm for the medium (detail G). Don't forget that some edges are cut at an angle of 80° and certain dimensions must be modified if using the 19 mm thick MDF. Let common sense guide you.

1.3.4 Assembly

For many years, I've been using 4 pin round Speakon© plugs instead of the traditional banana plugs. They have a lot of advantages (reliability, low resistance, protection, locking) and are inexpensive. For the connections, a four- wire cable is recommended. If the passive filter is mounted inside the enclosure, pins 1+ and 1- will be used for the low pass filter, pins 2+ and 2- will be connected to the high pass filter. For active filtering, pins 1+ and 1- will be connected to the SP1280, pins 2+ and 2- to the 140-15D.

The internal damping of an enclosure is always very important and raises many questions. What material should be used? Fiberglass, natural wool, foam, felt... and what quantity? What density?

It is not an exact science; however some general principles can be drawn:

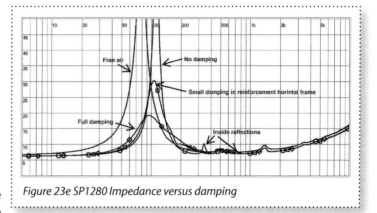

Figure 23e SP1280 Impedance versus damping

- Back wave velocity is maximum half way and null at wall position.
- For a vented high-pass enclosure, the airflow should be preserved for the vent to operate correctly. A compromise has to be found.
- Damping increases the apparent internal volume according to the stuffing material density that is used.

To illustrate the need for acoustical damping, I measured the SP1280 impedance versus frequency in the sealed high-pass enclosure with just 8 mm-thick felt on the walls (No damping fig. n°23e). Compared to the free air curve, two additional anomalies can be noticed, one at 340 Hz and to a lesser extend at 750 Hz. It may be a reflection of the back waves in the enclosure. With a velocity of 344 m/s, the half wave length at 340 Hz is about 50 cm. The greatest internal length inside the enclosure is 41.5 cm (diagonal). The 8.5 cm difference can be explained by the position of the SP1280 in the box and the horizontal reinforcement frame. To check this, I just put the small white damping stuff similar to fiberglass (visible in white picture n°18) in the reinforcement frame opening, which is almost halfway up the height of the enclosure. You can see that both anomalies are completely erased. As there is no other damping, the resonance frequency does not change much. It should be noted that the felt on the walls does not damp internal resonances but decreases micro vibration transmission to the box.

The impedance curve in light gray shows the full damped sealed high-pass enclosure as pictured on photos n°17 and n°18. It is noticeable that the resonance frequency is lowered by the apparent increase in volume and that the maximum impedance is considerably lower at around 20 Ω. I put a small sheet of 20 Kg/m³ fiberglass, maintained in position with low density polyester batting, in the middle of the top chamber where the velocity of the back waves are maximum. I've used good quality male banana plugs cut at one end and soldered to the cable for the 140-15D interconnections. A sheet of the same fiberglass was put in the lower compartment where it is maintained by the driver. I crimped 1/4" solderless terminals for the SP1280 interconnections. All these damping materials are likely to change in the future and everyone is invited to explore that field, but basic principles should always be kept in mind.

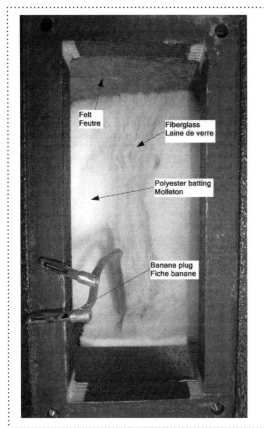
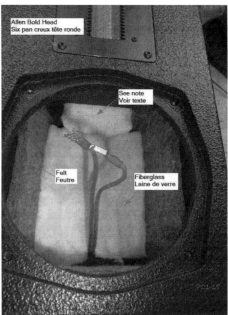

Picture n° 17 and n° 18

1.4   Measurements

Now's the time for measurements. But before that, let's talk about the problem of the sound space. All the simulations above are valid in half space 2π steradians. That is to say, as if the enclosure were embedded in an infinite and perfectly reverberant wall. In a perfectly anechoic soundproof room, bass propagate in a full space (4π steradians) because the wavelength is much larger (3.44 m at 100 Hz) than the front face. While the frequency increases, the wavelength decreases accordingly, until the dimensions of the front face are sufficient for the radiation to narrow to half space with a theoretical gain of 6 dB. It is easy to understand that, on the one hand, this transition is not as smooth as we would like it to be and, on the other hand, it is necessary to take that effect into account for the crossover. It is sometimes called baffle diffraction step. LEAP 5 not only simulates this kind of diffraction, but also approximates the diffraction edges based on driver location on the baffle. Of course, these simulations use simplified models but they remain useful to apprehend this type of problem. Figure n°24e shows the simulation of the sealed high-pass enclosure of figure n°11e but this time in full space.

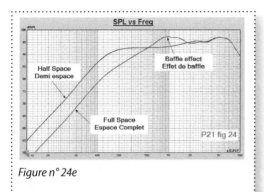

Figure n° 24e

And that changes everything. At 200 Hz, we notice the 6 dB loss of the full space radiation. At 500 Hz, the wavelength is short enough and the response reaches the level of the half space curve. Above 500 Hz, the baffle behaves like a reflector, bringing a maximum gain of 3 dB around 1 kHz. At 1500 Hz, the directivity of the SP1280 increases and the baffle effect fades out completely at 1900 Hz. LEAP 5 does not use finite elements for the calculations but simplified models, so these simulations are only rough estimates but the effect nevertheless does exist. As we normally listen to a loudspeaker in a variable semi-reverberant room we can imagine the difficulty to design a suitable filter. Picture n°19 shows the loudspeaker positioned for measure-

ments in the PHL AUDIO soundproof room. The microphone points exactly to the middle of the enclosure (between the medium and the tweeter) at 1.38 m from the front face. The AES panel, on the right of the picture, is put in a way to not interfere at all with measurements. The box on the rear left (box recommended by AES now) couldn't be removed but the interferences were acceptable and not significant for what I wanted to measure. I used a sample rate of 96 Ks/sec with a precision of 24 bit. The microphone has a useful bandwidth of 40 kHz.

Picture n° 19

1.4.1   SP1280

To ease the readability, I selected a horizontal scale from 200 Hz to 10 kHz and a 5 dB vertical scale. Response below 200 Hz is in conformity with the simulations and will be analyzed later on, in the chapter devoted to the active filtering. The response above 10 kHz is useless with a cutoff frequency in the 2 kHz range. Figure n°25e shows the SPL versus frequency of the SP1280 without any smoothing from 0° to 50° on the left by 10° steps. Measurements have also been taken on the right side but as the enclosure is symmetrical, they look identical.

Let's start with the on-axis response. There's a small dip (A) that has already been noticed on the free air impedance curve (figure n °5e). I can't think of any satisfactory explanation nor get the faintest one from the manufacturer. Anyway, it is of no real audible consequence from my point of view, because of the low amplitude and high Q. In the same way, the second anomaly, (B) also visible on the

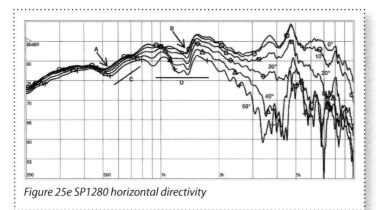

Figure 25e SP1280 horizontal directivity

impedance curve, is due to an interaction with the phase plug. Here too, the high Q and moderate amplitude make it inaudible. When the same driver was tested with a dust cap, the anomaly pointed to in B disappeared, but the area between 1.8 kHz and 3 kHz was of much lower amplitude. I've listened to both types of drivers for several months with adequate filtering and I've always preferred the sound of the one with the phase plug. Besides, we notice a first cone break-up at 3.7 kHz and a second one more pronounced (4 dB) at 4.8 kHz. This curve may seem jagged at this fine scale but the response of this driver is excellent. The horizontal directivity is minimal up to 2.8 kHz and very good up to 4,5 kHz from 0° to 30°. Area C will need some filtering to flatten the response and area D is where we should select the cut-off frequency.

The vertical directivity beneath the loudspeaker is shown in figure n°26e. For a better comparison, the same scale has been kept for each curve. The similarity with figure n°25e is not surprising, considering the location of the driver on the front face. The small differences, especially those visible above 3 kHz, are due to the differences in distance between the driver and the lower edge or the vertical sides and the angular degrees (90° instead of 80° for the vertical sides).

Figure n°27e shows the vertical directivity above the loudspeaker. The longer distance between the transducer and the upper edge causes some small undulations (550 Hz, 750 Hz...) but the amplitude excursions are moderate and should not be audible. From 1.5 kHz, the same kind of slopes as for the horizontal directivity can be observed.

The on-axis frequency and the polar response of this driver in this enclosure are really excellent. The trapezoidal shape, the small size and the driver recess certainly help. The harmonic distortion of the SP1280 (fig-

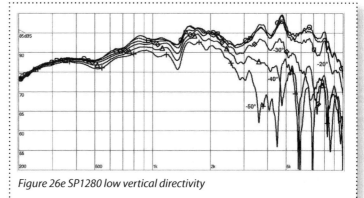

Figure 26e SP1280 low vertical directivity

ure n°28e) is very low, well below 0.1% in the useful bandwidth and almost flat from 150 Hz to 3.5 kHz. We can see that the third harmonic prevails, probably the consequence of an adequate flux ring. The small specific rise to 0.1% at 550 Hz corresponds to the anomaly on the impedance and amplitude

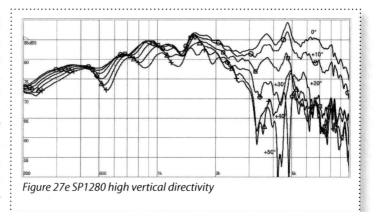

Figure 27e SP1280 high vertical directivity

curves, which has already been noticed.

1.4.2   140-15D

The 140-15D ribbon tweeter and the SP1280 medium driver were both measured under identical conditions. The left and right horizontal directivity are the same due to the driver position on the front plate. Figure n°29e shows the left horizontal directivity from 200 Hz to 40 kHz. The distance

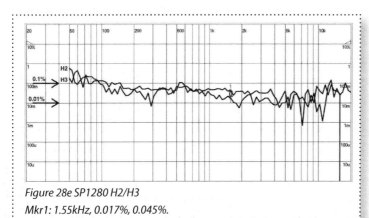

Figure 28e SP1280 H2/H3
Mkr1: 1.55kHz, 0.017%, 0.045%.

between foam pads remained at 50 mm during the whole measurement session.

It is exceptional to find a ribbon driver with such an extended and uniform horizontal polar response. On-axes the level at 40 kHz is the same as at 1500 Hz, at 30° it is still at -2 dB at 20 kHz. There's also a small depression in the 1 kHz area and a plateau up to 600 Hz. The chapter devoted to the numerical filter will provide more details about that, but let's briefly examine area "A". There is a 2.5 dB dip between 2 kHz and 3.5 kHz on-axes which is not replicated in the off-axes responses. It shows that this glitch does not come from the driver itself but from diffraction. So, if you use only the on-axes response to calculate a filter or a correction, you would be mistaken, because the compensation of the on-axis dip would result in an off- axes peak. Mr. Floyd E. Toole, an eminent acoustic expert, published, in January 2002, a

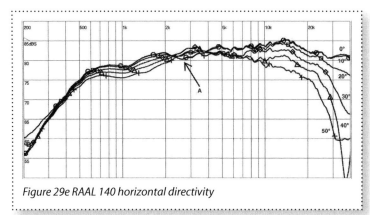

Figure 29e RAAL 140 horizontal directivity

three part paper[7] entitled "Loudspeakers and Rooms for Multichannel Audio Reproduction". I strongly recommend that you read it. Within the framework of this project, his article represents a very relevant digest.

The figures n°30e and n°31e show the low and high vertical directivity of the tweeter from 0° to 50°. For these measurements, the enclosure was laid flat on a side. I couldn't do the measurements otherwise. That's why the curves look a bit "grassy". A 1/6 octave smoothing would have removed this phenomenon entirely but I preferred to keep the curves untouched. I want also to point out that the microphone was not on the tweeter axis but in between the medium and tweeter, at half distance from each edge. In general, the vertical directivity for a planar or a ribbon tweeter is very pronounced but here, it is rather well controlled and uniform up to 30° with a regular decrease and without much irregularities up to 11 kHz. It is an excellent performance for a driver of this type. Of course, the responses at 40 ° and 50° from the axis are more irregular.

1.5 Passive crossover

Although the objective of the project is actually to develop a multi amplified DSP controlled loudspeaker system, I wondered whether it were possible to design a passive filter for the P21 mid-high satellite loudspeaker with acceptable performance. Various measurements (distortion and polar responses) show clearly that the cut-off frequency must range between 1 kHz to 2 kHz. The slope and the type of filter to be used have to be defined. Let us be clear from the start: when

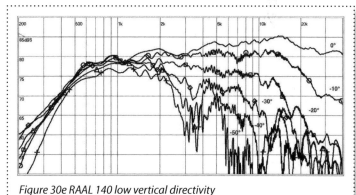

Figure 30e RAAL 140 low vertical directivity

7   http://www.infinitysystems.com/home/technology/technology_whitepapers.aspx?Language =ENG&Country=US&Region=USA

we speak about slope it is the overall slope including the response of the driver in the enclosure, plus the electrical behavior of the filter. This is valid for whatever type of filter, passive or active, analog or digital. In short, accurate measurements are required to design a good filter. I will talk about this

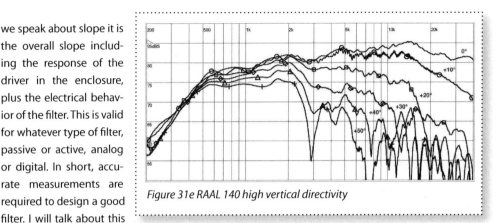

Figure 31e RAAL 140 high vertical directivity

topic in a chapter devoted to the DSP and I will give you some clues about affordable measurement software.

In audio, the most often used slopes vary from first to fourth order i.e. from 6 dB to 24 dB per octave. The usual types are Bessel, Butterworth and Linkwitz-Riley. Note that the later is characterized by cascading two even order Butterworth filters. The Legendre, Cauer and elliptic types are not so commonly used apart from the Neville-Thiele type which is an elliptic variant.

In spite of its excellent transient response, I have not considered a first order filter for three main reasons:

- The increased distortion of the tweeter below 1 kHz;
- The 6 dB/octave slope which will imply a substantial output from both drivers over a too large area which consequently will degrade the polar response;
- The impossibility of respecting the overall slope considering the driver responses.

Second order topology is difficult to calculate in a passive filter because it often results in a first order electrical filter, thus providing very little latitude for additional corrections such as diffraction. The overlap zone will still remain quite large near the cut-off frequency.

Third order filters are quite interesting; they will often result in a second order electrical topology, allowing some additional corrections with a minimum of components. Mr. Jean-Michel Le Cleac'h has developed an improved transient response version at the expense of a specific shift of the driver acoustic centers. Unfortunately, the choice of a flat front face makes it impossible to use this kind of filter with a passive crossover but I will pitch on it for the DSP controlled system.

I must admit that I'm fond of fourth order filters, more specifically those of the Linkwitz-Riley variety. The transient response is decent enough, the overlap zone is minimal, the - 6 dB at the cut-off frequency makes an interesting steep initial slope and the 180° phase shift puts both drivers in phase. What is most convenient with this kind of filter is that it gives large latitude in terms of additional correction possibilities. But a second to third order electrical filter is sufficient and you can use one of the inductors to correct the step diffraction.

The cut-off frequency will always result in a compromise. From the SP1280 point of view, and given

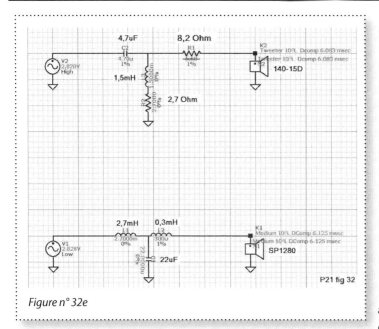

Figure n° 32e

the small glitch at about 1500 Hz, it would be better to place it towards 1 kHz or towards 2 kHz if necessary. From the 140-15D point of view, with the regular increase in distortion by third harmonic below 2 kHz and the light downward slope it would be better to chose it around 2 kHz. That's the frequency I selected as a starting point. Once the initial frequency was chosen, I hesitated between a third order Butterworth or a fourth order Linkwitz-Riley type. So, I computed all the corrections needed for both types in my BSS FDS 366T and listened to the loudspeakers, switching from one type to the other and after several days, I eventually adopted the forth order Linkwitz-Riley crossover.

Once again, I used LEAP 5 to home in on the filters values. In order to achieve accurate results, it is necessary to proceed step by step. After having imported the SP1280 averaged complex frequencies and the impedance curves, I started by correcting the baffle step diffraction. Then, I gave a fourth order low pass Linkwitz-Riley target with a 2 kHz cut-off frequency. Once convergence was obtained, I checked the sensitivity of the components, on the one hand, to determine those for which the tolerance is critical and, on the other hand, to try to eliminate those having a negligible effect on the result. I proceeded in the same way for the tweeter but with a high-pass filter. The next step was to combine both filters with the aim of obtaining the most linear response possible, only allowing the values of certain components to vary, to avoid abnormal results. This is a long process, the purpose of which is to lead to the best performance with the minimum of components. The best compromise was found while letting slip the cut-off frequency to 1800 Hz. These simulations must then be confronted with measurements and listening to adjust the final values.

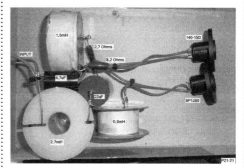

Picture n° 21

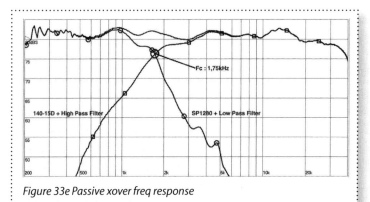

Figure 33e Passive xover freq response

1.5.1 Schematic

The final passive cross-over is shown in figure n°32. Only seven parts including two resistors are needed. The step diffraction is corrected by the 2.7 mH coil which is also a part of the third order electrical low-pass filter with the 0.3 mH coil and the 22 uF capacitor. The 8.2 Ω resistor reduces the level of the tweeter while the 4.7 uF, the 1.5 mH coil and the 2.7 Ω resistor build up the second order electrical high-pass filter.

1.5.2 Realization

I have used air core inductors with 15/10 mm wire section, 160V / 5% polypropylene capacitors and 10W / 5% non-inductive resistors. Picture n°21 shows a very compact implementation proposal. Note the position of the coils to reduce mutual coupling induction.

Once the crossover is assembled, the whole set can be put in a small wooden box. The inductors will be fixed in place with plastic collars or epoxy resin. You can fill the box with dry sand before closing it. It's advantageous to put the crossover near the amplifier, far from the vibrations of the enclosure, and to connect the low-pass and high-pass filter grounds separately to the amplifier(s).

It is still possible to include the crossover inside the enclosure, but since there is little room left in the enclosure and because of the inside sound pressure and micro vibrations, I am not in favor of this solution. An option is to split the low-pass and high-pass sections and put them into the two compartments. You can fix the 2.7 mH coil on the lower wall, the 0.3 mH on one of the vertical sides and the 22 uF capacitor near the Speakon socket. Try to put the coils as far as possible from the driver magnet. The high-pass section will find its place in the upper compartment. Fix all components carefully as you don't want them to generate additional parasitic noise.

1.5.3 Measurements

The measured acoustic response of the P21 satellite with the passive crossover (figure n°33e) makes it possible to check

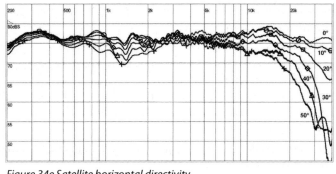

Figure 34e Satellite horizontal directivity

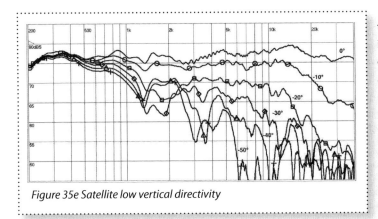

Figure 35e Satellite low vertical directivity

that the cut-off frequency is indeed at 1750 Hz / - 6 dB with stop band slopes approaching the theoretical 24 dB per octave attenuation. The summed frequency response curve has been taken at 10° left without any smoothing and in the same conditions as all preceding curves. The response is kept between ± 1.5 dB from 200 Hz to 30 kHz; this is really not so bad with only seven components for the filter. The horizontal directivity (figure n°34e) is excellent up to ± 30° off axis. These curves show some degradation between 1 kHz to 2 kHz at 40° and 50° that is logical considering the type of filter chosen. However, it remains limited and the response holds within ± 3.5 dB from 200 Hz to 15 kHz.

Vertical directivity (figures n°35e and n°36e) is quite smooth up to ± 20° from axis. A small dip around the cut-off frequency can be noticed which is due to the choices that have been made (type of filter and baffle size). At 30° it remains acceptable in spite of an increase in the dip already mentioned. At 40° and more, the situation is more complicated, but it should be remembered that a ± 20° smooth vertical beam seen from a distance of 2.5 m or more is already quite large.

The impedance curve of the passive version of the P21 satellite shown in figure n°37 never goes lower than 7.2 Ω (300 Hz) and should not be a problem for any amplifier. The phase remains contained within ± 30° maximum. It is perfectly possible to smooth out the 32 Ω impedance peak at 1.5 kHz by using an anti-resonant circuit, but that does not bring any benefits in terms of listening improvement.

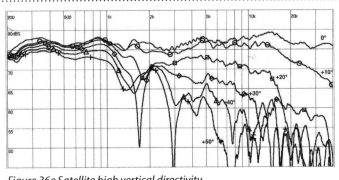

Figure 36e Satellite high vertical directivity

1.6   Stand

I've found a suitable stand[8] at an affordable price (photo n°1). The P21 loudspeaker weighs a lot

---

8   http://www.highland-audio.com/EN/Produits/Stand.html

more than what the "STAND" model originally was designed for. Therefore I reinforced the base with additional screws and replaced the upper wooden plate with an aluminum one. To increase stability I also added some lead sheets on the under-side. Your situation may be different but make

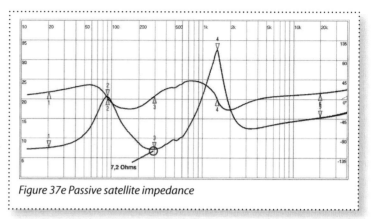

Figure 37e Passive satellite impedance

sure that the stand has enough stability for the weight of your satellites

1.7 Listening and Conclusion

It's always delicate for somebody who spent a lot of money and time bringing a project to a close, to objectively assess his own work. However, I've been listening to the P21 tri-amplified system for a year now and I've come to the conclusion that the sound quality is overwhelmingly good, indeed. And the comments of every credible person who has had the opportunity to listen to that system confirm my impressions. When associated with a pair (or more) of subwoofers with a cut off frequency of 100 to 140 Hz, dynamic is astonishing. The pinpoint sharp image reveals new and previously unsuspected sound subtleties, but unfortunately, sometimes dubious mixings too. It creates a remarkably open and spacious sound space but the image remains stable even when the listener moves. Surprisingly, sound wise the bi-amplified system (active subwoofer and passive mid-high satellites) is very close to the full tri-amplified one.

Of course, these speakers are quite expensive, especially the 140-15D that cost around 400 € each, but the result is worth the money. From a technical point of view, measurements are excellent. We have an on-axis flat frequency response holding within ±1.5 dB, a very low distortion and an efficiency of 87 dB/1 m/2.83V. From an acoustic point of view, the soundstage is gorgeous.

In Part II of this project article, I will present a subwoofer which will be adjustable to the room acoustics. I have created an email address dedicated to this project, should you wish to contact me: P21jcg@gmail.com.

Measurement of Thiele and Small parameters using the additional mass method

Briefly, the principle of this method is as follows: some characteristic values of the driver's impedance are measured in free air according to the diagram in the figure then a non-magnetic mass of precise value is added to the moving mass to cause a shift in resonance frequency of at least 20%, and the measurement is repeated.

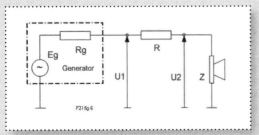

In practice, the line level output of a sound card, used as a generator, is connected to a low power amplifier and the U1 and U2 voltages are measured through the line level inputs. With current software, computer and sound cards, it has become very easy to extract these parameters in a quasi-automated way. But in order to obtain accurate results, it is necessary to take a certain number of precautions. Here are some important points to consider:

- *Use a slow (> 0.5 sec) sweep sine signal or better a step sine signal. The use of pseudo random noise like MLS (Maximum Length Sequence) even of a high order is not recommended because it does not load the driver enough in the lower frequencies.*

- *Pay attention to parasitic noise, the current flow in the driver is low and it should not be forgotten that the transducer behaves like a microphone too.*

- *Use a non-magnetic additional mass with a precisely known weight (for example with less than ±0.1 gr. for 10 gr.) that will allow a resonance frequency shift of about 20%.*

- *What series resistance value should we use? Mr. Thiele, in the early sixties, recommended a 1 kΩ value, therefore large enough compared to the driver impedance which made it possible to greatly simplify the calculations (quasi constant current). The problem is that high voltage is required to obtain enough current in the driver. Current measurement software suggests values in the 10 to 100 Ω range. With this order of magnitude, you neither work with constant current nor with constant voltage for a recommended standard. It is not a real problem but it is necessary to understand that parameters calculated this way could be slightly different from other methods without being inaccurate.*

- *Stay in the linear domain of the driver. Mr. Thiele and Mr. Small were relatively silent on this subject. Standard IEC 60208-5 is not of great help, it recommends use of a voltage or preferably a constant current low enough for the driver to remain in the linear range. However, it is also noted that impedance measurements are strongly influenced by the level which should neither be too high nor too low. A footnote suggests comparing the results at different levels to validate these parameters. It should be kept in mind that the more you increase the level in the linear range, the lower the resonance frequency and the Qts will be. If the ratio Fs/Qts remains almost constant, your measurement is valid.*

Low Frequencies in Rooms

Tom Nousaine

It's my opinion that people, by and large, suffer from many ill-conceived and mythological ideas about low frequencies in the home listening room. For example, it is generally thought that the dreaded room modes are the primary problem and need to be eliminated to make for good sound in rooms. In fact, most of the room EQ and subwoofer placement schemes are built on eliminating or alternately working hard to avoid exciting modes.

While this seems to be a straightforward approach it fails to recognize that modes in the room carry the energy. At any frequency, failing to excite a mode, leaves you with low sound pressure. For example play a 1000 Hz sine wave in your room and you can hear the sound pressure change when you move your head about even just a few inches. That's because the wavelength of the sound at 1000 Hz is only just a foot long. Of course, that's not really a problem because you will get the same effect at practically every location in the room. This makes it easy to deal with a kHz mode, for example you can equalize it and every listener will get the same effect. But remove the mode by eliminating a wall and then what happens? Sound pressure is reduced as the energy disappears into the atmosphere.

Of course, this gets more difficult to deal with because of the longer wavelengths at low frequencies below about 300 Hz and gets doubly difficult below 100 Hz where wavelengths are 10, 20 and even 50 or more feet long. That means that the variation in sound pressure at any listening position (seat to seat) can be more dramatic. But again you can reduce the effect by canceling the mode or equalizing the sound --- but the price is reduced sound pressure at frequencies where it's most difficult to produce it.

A related problem below 100 Hz is that there are few modes present. A typical room will have only 5 modes active below 100 Hz. For example in a 15 X 20 X 8 foot room you'll have modes of 35 and 70 Hz relative to the 15' width; 26 and 52 Hz relative to the length and at 66 Hz relative to the height. Now these are pretty good dimensions although the 3 modes that are at 52, 66, and 70 Hz may cause a slight build up around 65 Hz and the lack of anything in the 40 Hz region will probably

produce a "hole" in the sound pressure distribution at those frequencies. But let's say that you could magically "unexcite" all the modes; what happens then? No sound pressure at any location that isn't in the direct field of the speaker.

So let's examine this from a different perspective. As I see it even a good room (the one I just referenced was the most common room size found in a survey of 250 North American listening rooms conducted by a nationally distributed loudspeaker manufacturer) the "problem" isn't too many modes (which can be EQ'd) but not enough modes and/or an uneven distribution of them.

I began thinking about this more carefully when I received a device that was said to successfully "compensate" for low frequency unevenness in rooms. The apparatus was an electronic box that put out a noise signal and a microphone that sampled sound from a speaker and then applied a single frequency "cut" at a frequency that had a "peak." For example in a 11 X 10 X 8 foot room you'll have a 60 Hz build up in sound pressure because the modes all fall close to each other (48, 53, 66 Hz) and a single frequency cut EQ of the proper Q can improve the sound.

I figured that I could test the device by just finding a location, any location, where there was a sound pressure build up and see how well the device could handle it. But after trying and trying I just couldn't find any location, listening and otherwise, where I had such a condition in my primary listening room. Now I had carefully examined frequency response at several possible main listening positions/speaker locations when I made a DIY 5 Hz subwoofer with 8 15-inch Infinite Baffle drivers and knew that I had really good uniform bass response from the speaker/listener locations that I had chosen. But I hadn't mapped the entire room; only those listening positions and possible speaker locations.

Standing waves and the resultant modes occur between any two sets of parallel walls. After thinking about this for a while I simply just started counting the number of parallel walls in my "Great Room" listening room with an open stair case and a foyer and 2 wall openings. A quick count came up with arguably 7 sets of parallel walls, compared to just 3 sets in a normal rectangular room, all with differing distances and all would produce multiple well distributed modes below 100 Hz. Now that gives me smooth bass seat to seat but the price is that I have to have one hell of a lot of speaker displacement to energize the 7600 ft3 of room space to attain suitably high sound pressure levels.

After some more in-room measurements I also found a related issue with an extended listening space relative to subwoofer tuning - one that I had never heard being spoken of before. My primary listening space is a 6-foot sofa. And I wanted to have equal bass response at the three primary listening positions. My initial idea was to have the sofa at the west end of my room about 3 feet from the west wall. The subwoofer system is built into the basement and sound gets into the room through a 16 by 30-inch opening in the southwest corner of the room. I had originally figured this all out based on rectangular room thinking and had picked out a listening/viewing location that 'fit' putting a 6-foot screen at the east end of the room with the listening location at the west end which was relatively "close" to the subwoofer opening with magically smooth low frequency response at the primary listening location.

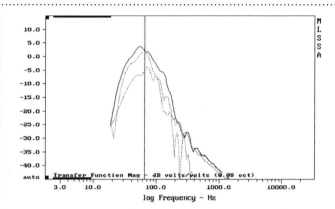

Figure 1: 10-inch corner subwoofer on West End of room. Smooth but level varies dramatically from seat to seat.

However what I found out was that even though I had found smooth frequency response over my 3 position seating area I couldn't get optimal level setting for every seat. See the graphs (Fig 1, 2) where there is a 10 dB level difference from one end of the couch to the other. That's because one seat is only 4 feet away from the subwoofer opening while the outside seat is 10 feet further away. So both seats cannot be accommodated with a single level setting. This is perhaps why multiple subwoofers have become so popular.

However, simply doubling your cost is seldom the best way of addressing a problem. And I wasn't going to double up on my 8 15-inch DIY Subwoofer in any case. So what to do? My solution was easy and simple too. Because of my smooth overall response I simply had to switch the sofa to the other end of the room where each seat was approximately the same distance from the sub corner.

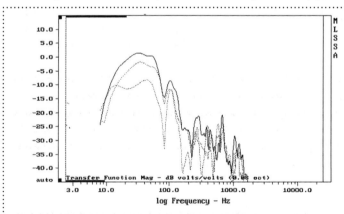

Figure 2: DIY Subwoofer in Corner on West End of Room: Same set, level smooth but one level setting can not serve three listeners optimally.

The charts (Fig 3, 4) show how this works with my DIY subwoofer and also with a Velodyne 10" powered unit. You can see some moderate modal notching which isn't occurring at the first seating location but the level

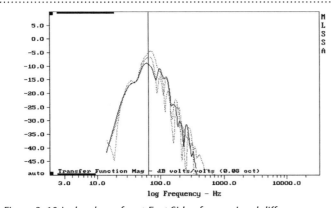

Figure 3: 10-inch subwoofer at East Side of room; Level differences are much smaller seat-to-seat.

difference seat-to-seat has been significantly moderated. So what kind of room is best for low frequencies? In my opinion the best room is built in the Great-Room architecture. One with multiple sets of parallel walls with a varied set of distances from each other. Is it perfect? No, of course not, but I think you start off with a greater possibility of optimization of bass response.

Well how about the mid-wall placement of 4 subwoofers in a rectangular room? I've done that experiment too. However that isn't perfect either and because it is built on the idea of not exciting modes even four of the subwoofers isn't quite enough to deliver the sound pressure capabilities of a single subwoofer mounted in a corner. That's because when you

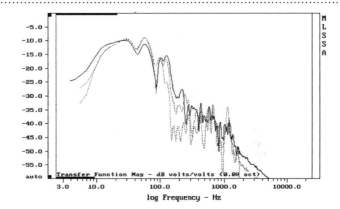

Figure 4: DIY Subwoofer at East Side of Room. Some moderate modal activity noticeable but levels are much more closely aligned.

fail to take advantage of modal activity you are limiting dynamic capability. Check out my website for more on this experiment. By the way, Earl Geddes suggests that multiple subwoofers even when randomly placed help even out bass response in smaller rooms.

(More on this subject can be found at the author's website www.nousaine.com and at www.gedlee.com - ed.)

The Distortion Magnifier

Bob Cordell

Measuring the very low distortion of modern amplifiers, preamplifiers and even op amps can be difficult with all but the most expensive test equipment. At the same time, sophisticated test functionality has become available to the hobbyist in the form of PC-based instrumentation that employs computer sound cards. This equipment, and reasonably-priced conventional analog test equipment, needs a means to improve its sensitivity and measurement floor.

The Distortion Magnifier (DM) described here provides ways of measuring very low levels of THD and IM distortions. These techniques go beyond the straightforward use of a THD or IM analyzer.

The Basic Idea

The Distortion Magnifier operates by subtracting the input of the amplifier under test from a scaled-down version of its output, leaving only the distortion products. One tenth of the amplifier output is then added back into the signal. The result is an output signal whose relative distortion has been magnified by a factor of ten. If only one percent of the amplifier's output is added back into the signal path, then the DM will have magnified the relative distortion level by a factor of 100. Key to this process is accurate level and phase matching of the source and amplifier output signal to be subtracted. The DM will work with any THD analyzer, spectrum analyzer or other type of measurement equipment, such as PC-sound card arrangements. The DM will enhance the dynamic range of the measuring equipment by 20 or 40 dB.

Because its operation is based on a signal subtraction process, the DM does not magnify the distortion or noise in the source test signal. This effectively increases the dynamic range of the source oscillator in the same way that it increases the dynamic range of the measurement equipment.

Features

- Coarse and fine amplitude and HF phase adjustment
- Selectable magnification of 1X (bypass) 10X or 100X
- Balanced differential inputs from DUT amplifier
- Reduction of relative distortion and noise inherent in the test source
- Measurements below −140 dB are achievable

The most straightforward use of the Distortion Magnifier (DM) is with a conventional THD analyzer. In this case, the DM is placed between the amplifier being tested (DUT) and the THD analyzer. By subtracting most of the sinusoidal test input applied to the DUT from the scaled-down output of the DUT, the distortion of the DUT is magnified by a controlled factor of 10 or 100.

A block diagram of the DM is shown in Figure 1. The DM is fed the source sinusoid and the output of the amplifier under test. The input and output signals of the amplifier under test are scaled to be of equal amplitude, adjusted for exact phase match, and subtracted. A selectable amount of the signal from the DUT is then added back to the result so that there is some known value of the fundamental for the subsequent THD analyzer to lock onto. This process results in a relative magnification of the distortion by a factor of 10 or 100.

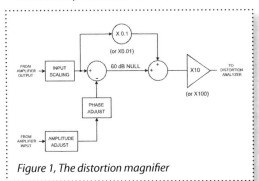

Figure 1, The distortion magnifier

If the THD analyzer normally has a residual measuring capability (measurement floor) of 0.001 percent, and if the DM is set to a magnification of 10X, then the combined instruments are capable of a measurement floor of 0.0001 percent as long as the distortion and noise contributed by the DM are less than that amount. In such an arrangement, when the THD analyzer is reading 0.001%, the actual distortion of the DUT is 0.0001%.

A typical front panel layout of the DM is shown in Figure 2. Balanced inputs are available for the signals from the DUT, while a single-ended input receives a copy of the source signal applied to the DUT. An additional input allows the injection of a known "distortion" test signal into the signal path for calibration purposes. A monitor output is provided that supplies a copy of the received DUT output signal attenuated by a factor of 10. Four potentiometers provide coarse and fine control of gain and phase matching for the nulling adjustment. A three-position switch allows selection of 10X or 100X distortion magnification, or the nulling mode. A second switch allows the gain in the signal path to be increased by a factor of 10 to restore the level of the fundamental to its nominal value when in the X100 magnification mode. A third switch allows the magnification process to be bypassed, effectively providing an X1 magnification function that is useful for reference purposes.

Amplitude and Phase Matching

The most fundamental operation in the DM is the creation of a null by subtracting the test signal input from a scaled-down version of the output of the DUT. The objective is to achieve a 60 dB (or better) null in this process. This requires accurate phase and amplitude matching, so both coarse and fine amplitude and phase adjustments are desirable. The DM is set up to accommodate DUT gains from about 10 to about 35.

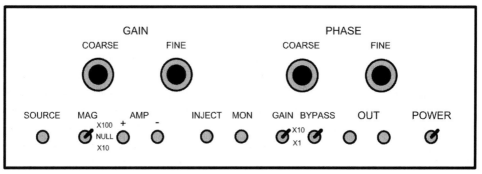

Figure 2: Example front panel layout for the DM

High-frequency phase matching is achieved with an adjustable single-pole LPF roll-off in the test signal input path. The LPF emulates the phase and delay characteristics of the DUT, at least in the neighborhood of the test signal frequencies and their harmonics. More complex phase matching networks could be used, but have thus far proved unnecessary.

The DM could also incorporate an adjustable low frequency phase matching network to compensate for the usual high-pass characteristic of an AC-coupled DUT (e.g., a 3-dB corner at, say 5 Hz) to enable more accurate nulls to be achieved when implementing low-frequency THD measurements. This has not been incorporated into the DM at this time.

The first-order phase compensation networks incorporated into the current DM are only an approximation to the phase match that is accurate at the given test frequency used when the null adjustment is made. If a distortion measurement is made at 20 kHz and then the measurement is to be carried out at 10 kHz, some re-adjustment of the null may be necessary.

The phase compensation approximation will also be less accurate for cancellation of harmonics of the test signal that are present in the sine wave source. Great improvement in immunity to sine wave source distortion will still be had, but for this reason it is still desirable to employ a low-distortion sine wave source for best results.

The Nulling Process

The nulling process is an iterative exercise between the phase and amplitude matching controls, but it actually converges quite quickly. Nulling is accomplished by monitoring the output of the DM with an AC voltmeter while the mode switch in the "null" position. In this position, no fraction of the DUT test signal is added back into the signal path. The fine amplitude and phase controls are set to their mid position and the coarse amplitude and phase controls are adjusted for a good minimum. Adjustment of the fine controls is then iterated until a null of at least 60 dB is achieved. A more advanced version of the DM would incorporate a built-in logarithmic amplitude display to make the nulling adjustment easier.

DUT Input Channel

The DUT input channel is shown in Figure 3. The balanced input from the DUT at J1A and J1B is applied to a differential amplifier (U1A) with a gain of 0.1 (the right side of dual RCA connectors will be referred to as the B connector). Most power amplifiers have a single-ended output, in which case the negative half of the balanced input is simply connected to the speaker return terminal at the amplifier. The balanced inputs can be swapped for power amplifiers that are inverting. The use of balanced inputs on the DM for receiving the output of the DUT is important even when the DUT is not a balanced amplifier. This is so because we want to reject any noise or distortion introduced by testing ground loops or stray magnetic fields. Input resistors R1 and R2 are 2-watt metal film types to minimize any heating-induced distortions.

A calibrated amount of "distortion" can be injected at J3A from a signal generator to check overall system distortion measurement accuracy and scaling. This signal will be attenuated to 1/1000 of the DUT output. So if you inject the same level as the DUT output, which is 100%, the 1/1000 attenuation should make for a "distortion" indication of 0,1%. If the DUT produces 20V RMS and you inject a "distortion" calibration signal of 2V RMS you should get an indication of 0,01%. The signal injected can be of any desired frequency and need not be related to the main test signal.

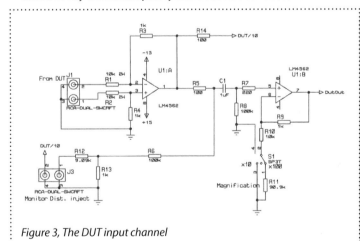

Figure 3, The DUT input channel

J3B provides a MONITOR output equal to 1/10 the level of the DUT. This output is also fed to the main output of the DM (Figure 5) when the DM is in BYPASS mode.

The single-ended and scaled DUT signal is then passed through near-unity-gain amplifier U1B. In the "null" mode, MAGNIFICATION switch S1 is in its center open position and the gain of this stage is exactly unity. Under this condition the gain and phase controls are adjusted for the deep null.

For measurements with a magnification factor of 10 or 100, the injection of a little bit of extra signal from the DUT is simply accomplished by increasing the gain of this stage ever so slightly to either 1.1 or to 1.01 by moving the swinger of S1 to a position that will engage a feedback shunting resistance (R10 or R10+R11) to work against feedback resistor R9.

Source Input Channel

The source input channel (J5, Figure 4) takes a copy of the signal that was applied to the input of the power amplifier DUT and adjusts its amplitude and phase to achieve a near-perfect null at the subsequent summer (which actually performs a subtraction).

Figure 4 shows the source input channel. It comprises a single op amp, U2A, that acts as both a gain control and a buffer. The non-inverting amplifier formed by U2A has a nominal gain of 1.3. This nominal gain is offset by R16 and R18 so that the nominal overall gain of the channel is approximately unity. This corresponds to a DUT with a gain of 20 (26 dB). Coarse gain pot RV1 allows the gain to be varied to accommodate power amplifier gains ranging from about 10 to 35. Fine gain control pot RV2 enables precise adjustment to achieve a deep null.

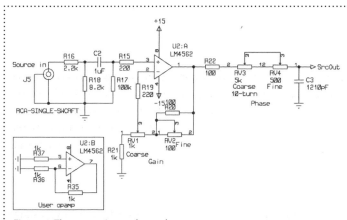

Figure 4, The source input channel

The buffered output of U2A goes to the phase-compensation low-pass filter. Potentiometers RV3 and RV4 provide the series resistance, while C3 provides the shunting capacitance. The single pole is nominally at about 48 kHz when the pots are centered. With both pots set to their minimum resistance, the pole is above 1 MHz. For most applications the lowest frequency available for the pole is unnecessarily low, but this can accommodate some vacuum tube amplifiers of limited bandwidth (for which a distortion magnifier would probably be unnecessary).

Summer and Output Amplifier

The summer and output amplifier circuit is shown in Figure 5. The summer is where the subtraction for the null takes place. Here the output of the source input channel is subtracted from the output of the DUT input channel. This circuit also provides selectable gain for the resulting difference signal for generating the output of the DM.

Differential amplifier U3B performs a subtraction of the DUT and source channel signals. It is arranged in such a way that the output is zero when the signal from the DUT channel is exactly twice the signal from the source channel (the factor of two here relates to the nominal gain of the DUT

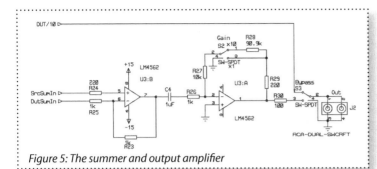

Figure 5: The summer and output amplifier

being 20).

The output amplifier is implemented by U3A. It provides a gain of either 10 or 100, corresponding to the magnification factors of 10 and 100. This gain keeps the amplitude of the fundamental (as seen by the THD analyzer) the same whether the DM is in the BYPASS mode (see S3) or in either of the magnification modes. The only thing that changes among these three modes is the effective amount of distortion magnification. The amount of fundamental at the output of the DM will remain at 1/10 the level at the output of the DUT.

Powering the DM

Figure 6 shows the details of the ±15V power supply for the DM. A conventional bridge rectifier arrangement provides raw DC that is regulated by LM317/337 IC regulators (U4, U5). The DM can be powered from a center-tapped power transformer rated at 24V RMS or more from end to end, but this is marginal. The DM can also be powered from a 20V AC wall transformer rated at 10VA or more. In this arrangement power pins J6 and J8 are both connected to one side of the wall transformer secondary and ground is connected to the other side.

Using the Distortion Magnifier

For spectral THD measurement, the distortion residual from a THD analyzer is fed to a spectrum analyzer. The spectrum analyzer separates out the noise and any remaining fundamental from the various distortion products, yielding a much more sensitive arrangement. Even greater sensitivity can be had with the use of the Distortion Magnifier circuit in such an arrangement, as illustrated in Figure 7.

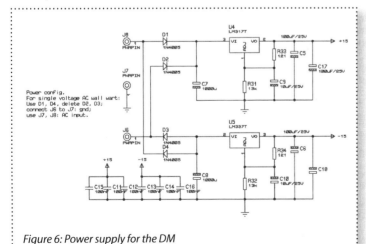

Figure 6: Power supply for the DM

Use of the DM allows a relative magnification of the distortion by a factor of 10 or 100. Note that distortion and noise in the sinusoidal source are also reduced in the same relative proportion as the magnification factor. As long as the DM is implemented with operational amplifiers with very low distortion (like the LM4562), the measurement floor of the resulting system is reduced by 20-40 dB. Given the ability of the spectrum analyzer to largely eliminate noise contributions, measurement floors of −140 dB or better are achievable.

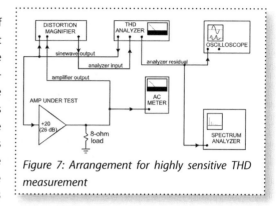

Figure 7: Arrangement for highly sensitive THD measurement

Use with Other Types of Distortion Tests

The Distortion Magnifier is also useful with more complex distortion tests such as twin-tone CCIF, SMPTE IM and DIM, where the test signal is not just a simple sinusoid. Many of these more complex tests rely on the use of a spectrum analyzer, where useable dynamic range is always at a premium. The DM increases the useable dynamic range in such arrangements by 20 or 40 dB. Indeed, since the spectrum analyzer does not require a fundamental to lock onto, the DM can be operated in its full null mode, where none of the test signal is added back into the signal path. In this case, its magnification of the useable dynamic range can be on the order of 60 dB. In this case the actual magnitude of each harmonic as seen on the SA must be compared to the known output level of the DUT after appropriate scaling. Applying a small signal of known amplitude to the injection input can be helpful here.

Twin tone IM distortion is a good example of the use of the DM with more complex tests. CCIF IM is measured by summing two high-frequency tones, such as 19 kHz and 20 kHz, and applying the result to the amplifier under test. The output of the amplifier is then observed with a spectrum analyzer. This measurement can also benefit in the same way from the use of the Distortion Magnifier.

Use with PC-based Test Equipment

Today's availability of PC-based instrumentation, often employing sound cards, has made many sophisticated measurement capabilities available at very low cost. Perhaps the best example of this is the spectrum analyzer function based on the FFT. In the past, conventional analog spectrum analyzers, like the HP 3580A, were very expensive.

PC-based instrumentation is limited in its performance capabilities by the quality of the sound card and the noisy environment in which it may reside. The Distortion Magnifier is thus a perfect complement to PC-based instrumentation. Not only does it magnify the distortion for better dynamic range, but it also provides a handy interface between the DUT and the input of the sound card.

If the DM is to be used with a sound card, it is a good idea to add an amplitude-limiting circuit to

the output of the DM so as not to overload or damage the sound card. This could be as simple as adding two pairs of 1N4148 diodes connected in opposing polarity across the DM output at J2. This will limit the output to less than about 1.4V peak. One must watch for distortion added by this arrangement, however. A greater number of diodes in series, or low-voltage Zener diodes can be used to achieve higher limiting thresholds if the sound card can accommodate higher levels. More sophisticated limiting circuits can be imagined as well.

The performance of the DM is essentially as good as the performance of its op amps and their associated circuits. This applies to both noise and distortion. The objective is for the DM to have better noise and distortion performance than the DUT. This is not difficult with today's excellent op amps. Indeed, very impressive performance of the DM is achieved using the LM4562 op amps.

As noted above, the noise-dominated performance can be largely improved by the use of a spectrum analyzer, especially if it is set to a very small measurement bandwidth.

Building the Distortion Magnifier

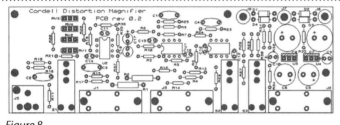

Figure 8

A PCB has been made available for the DM, which holds all the parts except the 4 potentiometers for the coarse and fine level and phase adjustments. The board layout is shown in Fig 8; building the unit using this board will virtually guarantee success. A bill of materials is at the end of the article and gives the Mouser order numbers for the switches and RCA jacks that fit the board. A semi-kit containing the PCB and some selected parts is available from Pilgham Audio, check their website at www.pilghamaudio.com.

Conclusion

The Distortion Magnifier should prove to be a useful addition to the bench of anyone doing serious distortion measurements. It can dramatically increase the dynamic range and sensitivity of both conventional analog distortion analyzers and those based on a PC-soundcard combination.

Bill of materials

Resistors: (all ¼ W metal film 1% except
where noted)
R1,R2: 10k 2W
R3,R4,R9,R13,R21,R23,R25,R26,R35-R37: 1k

R5,R14,R20,R22,R30: 100 ohms
R6,R8,R17: 100k
R7,R15,R19,R24,R29: 221 ohms
R10,R27: 10k

R11,R28: 90.9k
R12: 9.09k
R16: 2.21k
R18: 8.25k
R31,R32: 13k
R33,R34: 121 ohms

Opamps:
U1-U3: LM4562, DIL08
Voltage regulators:
U4: LM317T, TO220
U5: LM337T, TO220

Potentiometers:
RV1: 1k; RV2: 100 ohms; RV3: 5k (preferably multiturn); RV4: 500 ohms

Switches:
S1: SP3T, NKK M2T13SA5G40-RO, Mouser # 633-M2T13SA5G40-RO
S2,S3: SPDT, NKK M2T12SA5G40-RO, Mouser # 633-M2T12SA5G40-RO

Capacitors:
C1,C2,C4: 1uF, 50V, film
C3: 1210pF, 50V, 5% film
C5,C6: 100uF/25V electrolytic
C7,C8: 1000u/35V electrolytic
C9,C10: 10uF/25V electrolytic
C11-C16: 100nF, 50V decoupling
C17,C18: 100uF/25V electrolytic

Connectors:
J1-J3: RCA dual, switchcraft PJRAS2X1S01X, Mouser # 502-PJRAS2XS01X
J5: RCA single, switchcraft PJRAS1X1S02X, Mouser # 502-PJRAS1X1S02X)

Diodes:
D1-D4: 1N4005 or eq.

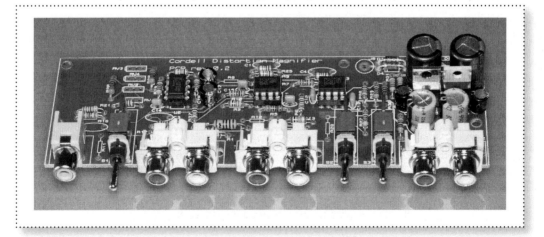

On the leakage inductance in audio transformers

Pierre Touzelet

Introduction

The leakage inductance of a transformer is the result of the magnetic flux that does not link primary and secondary windings. It appears in series with the primary winding and combined with the stray capacitance across the primary winding, generates a low pass filter responsible of the transformer performance at high frequencies. An extended response towards high frequencies requires both smallest possible leakage inductance and stray capacitance. The leakage inductance reduction is generally achieved using split and overlapped primary and secondary windings whereas capacitance reduction if achieved using thick insulation between winding layers. This paper deals only with the leakage inductance. It shows how, with reasonable assumptions on the transformer topology, simple analytical analysis can be performed, allowing for the evaluation and minimization of leakage inductance with acceptable accuracy.

Schematic diagram and notations

Figure 1 shows a transformer having a cylindrical symmetry, its primary and secondary windings split into q blocks and overlapped in such a way that the primary and secondary blocks k, $\forall k = 1, 2, \cdots, q$ are wound close to each other.

Notations are defined as follows:

zz'	Axis of symmetry
$r_{p_{k1}}$	Internal radius of the primary block k
$r_{p_{k2}}$	External radius of the primary block k
$r_{s_{k1}}$	Internal radius of the secondary block k
$r_{s_{k2}}$	External radius of the secondary block k
n_p	Number of turns of the primary winding
n_{p_k}	Number of turns of the primary winding block k
n_s	Number of turns of the secondary winding

n_{s_k} Number of turns of the secondary winding block k

i_p Primary current

i_s Secondary current

L   Windings width

aa'bb'   Closed path

$H(r)$ Magnetic field intensity along the bb' line

Assumptions

In order to allow analytical analysis, we assume the following reasonable assumptions, which do not deviate significantly from those of a real transformer.

The transformer has a cylindrical symmetry around the zz' axis.

Along lines such as bb' the magnetic field intensity, responsible of the magnetic flux leakage, is constant as a result of the cylindrical symmetry.

Along the line aa', located between the magnetic core and the internal diameter of the primary block 1, the magnetic field intensity, responsible of the magnetic flux leakage, is equal to zero.

Along lines such as ab and a'b' located between the magnetic core and the primary and secondary winding blocks, the magnetic field intensity, responsible of the magnetic flux leakage, is perpendicular to these lines.

Proposed solution

Ampere's theorem applied along the closed path aa'bb' gives, according to the above assumptions:

$$\oint_{aa'bb'} Hdl = H(r)L = \sum (enclosed \cdot current)$$

From transformer theory we know:

$$n_p i_p = -n_s i_s$$

Let:

$$\sum_{j=1}^{q} (r_{Pj2} - r_{Pj1}) = e_p \text{ Thickness of the primary winding without splitting}$$

$$\sum_{j=1}^{q} (r_{sj2} - r_{sj1}) = e_s \text{ Thickness of the secondary winding without splitting}$$

$$\left(r_{P_{k2}} - r_{P_{k1}}\right) = e_{P_k} \qquad \text{Thickness of the primary winding block } k$$

$$\left(r_{s_{k2}} - r_{s_{k1}}\right) = e_{s_k} \qquad \text{Thickness of the secondary winding block } k$$

$$\left(r_{P_{k2}} - r_{s_{k1}}\right) = e_{g_{k,k}} \qquad \text{Thickness of the gap between blocks } k$$

$$\left(r_{s_{k2}} - r_{P_{(k+1)1}}\right) = e_{g_{k,k+1}} \qquad \text{Thickness of the gap between blocks } k \text{ and } k+1$$

$$\frac{\sum_{i=1}^{k}\left(r_{P_{i2}} - r_{P_{i1}}\right)}{\sum_{j=1}^{q}\left(r_{P_{j2}} - r_{P_{j1}}\right)} = X_{P_k} \quad \text{Thickness ratio involved with the first } k \text{ primary blocks}$$

$$\frac{\sum_{i=1}^{k}\left(r_{s_{i2}} - r_{s_{i1}}\right)}{\sum_{j=1}^{q}\left(r_{s_{j2}} - r_{s_{j1}}\right)} = X_{s_k} \qquad \text{Thickness ratio involved with the first } k \text{ secondary blocks}$$

With these notations, we have:

Boundary conditions

$$X_{P_0} = X_{s_0} = 0$$

$$X_{P_q} = X_{s_q} = 1$$

Field intensity $H\left(r\right)$

If $r_{s_{(k-1)2}} < r < r_{P_{k1}}$

$$H\left(r\right) = \frac{n_p i_p}{L}\left[X_{P_{k-1}} - X_{s_{k-1}}\right]$$

If $r_{P_{k1}} \le r \le r_{P_{k2}}$

$$H\left(r\right) = \frac{n_p i_p}{L}\left[X_{P_{k-1}} - X_{s_{k-1}} + \frac{\left(r - r_{P_{k1}}\right)}{e_p}\right]$$

If $r_{Pk2} < r < r_{sk1}$

$$H(r) = \frac{n_p i_p}{L}\left[x_{Pk} - x_{sk} + \frac{e_{psk}}{e_s}\right]$$

If $r_{sk1} \leq r \leq r_{sk2}$

$$H(r) = \frac{n_p i_p}{L}\left[x_{Pk} - x_{sk} + \frac{(r_{sk2} - r)}{e_s}\right]$$

The stored magnetic energy, due to the magnetic field intensity, responsible of the flux leakage and leakage inductance, for the primary and secondary blocks k, is:

$$E_k = \frac{1}{2}l_{lk} i_p^2 = \frac{\mu_0}{2}\iiint_{V_k} H^2 dv$$

Where:

μ_0 Free space permeability

V_k Volume occupied by the primary and secondary blocks k, the gap

         between blocks $k-1$ and k, the gap between blocks k

l_{lk} Leakage inductance due to blocks k

According to the cylindrical symmetry, we have:

$$E_k = \mu_0 \pi L\left[\int_{r_{s(k-1)2}}^{r_{Pk1}} H^2 rdr + \int_{r_{Pk1}}^{r_{Pk2}} H^2 rdr + \int_{r_{Pk2}}^{r_{sk1}} H^2 rdr + \int_{r_{sk1}}^{r_{sk2}} H^2 rdr\right] = E_{k_1} + E_{k2} + E_{k_3} + E_{k4}$$

We have for E_{k_1}

$$E_{k_1} = \frac{\mu_0 \pi n_p^2 i_p^2}{L}\int_{r_{s(k-1)2}}^{r_{Pk1}} \left(x_{Pk-1} - x_{sk-1}\right)^2 rdr = \frac{\mu_0 \pi n_p^2 i_p^2}{2L}\left(x_{Pk-1} - x_{sk-1}\right)^2 \left(r_{Pk1}^2 - r_{s(k-1)2}^2\right)$$

E_{k_1} is the magnetic energy stored into the gap between blocks $k-1$ and k. This energy can be made equal to zero if $x_{Pk} = x_{sk} = x_k\ \forall k = 1, 2, \cdots, q$ minimizing E_k and hence, l_{lk}. This condition is equivalent to stating that primary and secondary blocks k have the same magnetomotive force absolute values.

Assuming that the transformer configuration meets this simple first minimizing condition, the integration process is strongly simplified and gives:

$$E_{k_2} = \frac{\mu_0 \pi n_p^2 i_p^2}{Le_p^2} \int_{r_{Pk1}}^{r_{Pk2}} \left(r - r_{Pk1}\right)^2 rdr = \frac{\mu_0 \pi n_p^2 i_p^2 \left(r_{Pk2} - r_{Pk1}\right)^3}{12Le_p^2}\left(3r_{Pk2} + r_{Pk1}\right)$$

$$E_{k_3} = \frac{\mu_0 \pi n_p^2 i_p^2 \left(x_{Pk} - x_{s_{k-1}}\right)^2}{L} \int_{r_{Pk2}}^{r_{sk1}} rdr = \frac{\mu_0 \pi n_p^2 i_p^2 \left(r_{s_{i2}} - r_{s_{i1}}\right)\left(r_{Pk2} - r_{sk1}\right)}{2L\left[\sum\limits_{k=1}^{q}\left(r_{s_{k2}} - r_{s_{k1}}\right)\right]^2}\left(r_{Pk2} + r_{sk1}\right)$$

$$E_{k_4} = \frac{\mu_0 \pi n_p^2 i_p^2}{Le_s^2} \int_{r_{sk1}}^{r_{sk2}} \left(r_{sk2} - r\right)^2 rdr = \frac{\mu_0 \pi n_p^2 i_p^2 \left(r_{sk2} - r_{sk1}\right)^3}{12Le_s^2}\left(3r_{sk1} + r_{sk2}\right)$$

Summing, we get:

For the leakage inductance, related to primary winding of blocks k

$$l_{l_k} = \frac{\mu_0 \pi n_p^2}{L}\left[\frac{\left(r_{Pk2} - r_{Pk1}\right)^3\left(3r_{Pk2} + r_{Pk1}\right)}{6e_p^2} + \frac{\left(r_{sk2} - r_{sk1}\right)^3\left(3r_{sk1} + r_{sk2}\right)}{6e_s^2} + \frac{\left(r_{s_{i2}} - r_{s_{i1}}\right)^2\left(r_{Pk2} - r_{sk1}\right)\left(r_{Pk2} + r_{sk1}\right)}{e_s^2}\right]$$

For the global leakage inductance related to the primary winding

$$l_l = \sum\limits_{k=1}^{q} l_{l_k} = \frac{\mu_0 \pi n_p^2}{L}\left[\sum\limits_{k=1}^{q}\left[A_k\left(3r_{Pk2} + r_{Pk1}\right)\right] + \sum\limits_{k=1}^{q}\left[B_k\left(3r_{sk1} + r_{sk2}\right)\right] + \sum\limits_{k=1}^{q}\left[C_k\left(r_{Pk2} + r_{sk1}\right)\right]\right]$$

Where:

$$A_k = \frac{\left(r_{Pk2} - r_{Pk1}\right)^3}{6e_p^2} = \frac{e_{Pk}^3}{6e_p^2}$$

$$B_k = \frac{\left(r_{sk2} - r_{sk1}\right)^3}{6e_s^2} = \frac{e_{sk}^3}{6e_s^2}$$

$$C_k = \frac{\left(r_{sk2} - r_{sk1}\right)^2\left(r_{Pk2} - r_{sk1}\right)}{e_s^2} = \frac{e_{sk}^2 e_{gk,k}}{e_s^2}$$

Additional simplifications

To go further, the following reasonable additional simplifications can be made.

Let:

d_m Average diameter of the overlapped primary and secondary windings

$P = \pi d_m$ Average perimeter of the overlapped primary and secondary windings

$e_{g_{k,k}} = e_g$ Constant gap thickness between primary and secondary blocks k

With these additional assumptions, we have approximately:

$$\forall k = 1, 2, \cdots, q$$

$$3r_{p_{k2}} + r_{p_{k1}} \approx 2d_m$$

$$3r_{s_{k1}} + r_{p_{k2}} \approx 2d_m$$

$$r_{p_{k2}} + r_{s_{k1}} \approx d_m$$

Substituting into the global leakage inductance formula, we get:

$$l_l = \frac{\mu_0 n_p^2 P}{L} \left[\frac{1}{3e_p^2} \sum_{k=1}^{q} e_{p_k}^3 + \frac{1}{3e_s^2} \sum_{k=1}^{q} e_{s_k}^3 + \frac{e_g}{e_s^2} \sum_{k=1}^{q} e_{s_k}^2 \right]$$

We have also:

$$e_{p_k} = \left(x_{p_k} - x_{p_{k-1}} \right) e_p$$

$$e_{s_k} = \left(x_{s_k} - x_{s_{k-1}} \right) e_s$$

With:

$$x_{p_k} = x_{s_k} = x_k$$

Finally, the global leakage inductance formula becomes:

$$l_l = \frac{\mu_0 n_p^2 P}{L} \left[\frac{(e_p + e_s)}{3} \sum_{k=1}^{q} \left(x_k - x_{k-1} \right)^3 + e_g \sum_{k=1}^{q} \left(x_k - x_{k-1} \right)^2 \right]$$

Splitting optimization of primary and secondary windings

Let:

$$\left(x_k - x_{k-1} \right) = \delta_k$$

The problem to solve is the following:

Find the values of variables δ_k $\forall k = 1, 2, \cdots, q$ which minimize the function:

$$f(\delta_1, \delta_2, \cdots, \delta_q) = \frac{\mu_0 n_p^2 P}{L} \left[\frac{(e_p + e_s)}{3} \sum_{k=1}^{q} \delta_k^3 + e_g \sum_{k=1}^{q} \delta_k^2 \right]$$

These variables being linked by the relationship

$$g(\delta_1, \delta_2, \cdots, \delta_q) = 1 - \sum_{k=1}^{q} \delta_k = 0$$

This problem is a standard problem of linked extremums which is solved using Lagrange's multipliers method.

Let: λ be a Lagrange multiplier. We solve for δ_k the following system:

$$\frac{\partial (f + \lambda g)}{\partial x_k} = \frac{\mu_0 n_p^2 P}{L} \left[(e_p + e_s) \delta_k^2 + 2 e_g \delta_k \right] - \lambda = 0$$

We get:

$$\delta_k = \frac{e_g + \sqrt{e_g^2 + \dfrac{(e_p + e_s)}{\alpha} \lambda}}{(e_p + e_s)}$$

with $\quad \alpha = \dfrac{\mu_0 n_p^2 P}{L}$

To determine λ, we use the relationship

$$g(\delta_1, \delta_2, \cdots, \delta_q) = 1 - \sum_{k=1}^{q} \delta_k = 0$$

We find:

$$\lambda = \frac{\alpha \left[\left(\dfrac{(e_p + e_s)}{q} - e_g \right)^2 - e_g^2 \right]}{(e_p + e_s)}$$

Substituting the above value for λ in δ_k gives

$$\delta_k = \frac{1}{q}$$

Assuming, that the transformer configuration meets this simple second minimizing condition, we get for the general formula of the minimized leakage inductance, related to the primary winding:

$$l_l = \frac{\mu_0 n_p^2 P}{q^2 L} \left[\frac{\left(e_p + e_s\right)}{3} + q e_g \right]$$

Comments

The above analysis is only possible because we used an idealized transformer topology. Fortunately real transformers do not differ significantly, and results obtained provide a leakage inductance value with an acceptable accuracy, for most of the transformer projects in audio applications.

The proposed analytical leakage inductance formula is similar to the one often found in the literature. But most of the time, the formula is given without any mention about the conditions it requires to make it valid.

The above results apply only when primary and secondary are split into the same number of blocks. This point is often forgotten despite it is needed to comply with the first minimizing condition.

Primary and secondary blocks can be arranged in series or in parallel without changing the results.

The primary and secondary blocks, of the same index, can be interchanged without changing the results. This remark is of importance because, used rationally, as shown in figure 2, it allows the merging of blocks of the same category, reducing, as a result, the effective number of windings and interfaces. Despite the fact that both arrangements are strictly equivalent, the second one is often preferred by manufacturers because of its greater simplicity.

A transformer where the primary and secondary windings are not split $\left(q = 1\right)$ shows the leakage inductance:

$$l_l = \frac{\mu_0 n_p^2 P}{L} \left[\frac{\left(e_p + e_s\right)}{3} + e_g \right]$$

The same transformer with split and overlapped primary and secondary windings shows the leakage inductance:

$$l_l = \frac{\mu_0 n_p^2 P}{q^2 L} \left[\frac{\left(e_p + e_s\right)}{3} + q e_g \right]$$

It can be seen that the splitting and overlapping of the primary and the secondary windings provide a reduction factor, for the leakage inductance, which is:

$$\eta = \frac{1}{q^2} \left[\frac{\dfrac{e_p + e_s}{3} + e_g}{\dfrac{e_p + e_s}{3} + qe_g} \right]$$

demonstrating, despite the resulting winding complexity, all the benefits of this technique.

Conclusions

In this paper, I have shown that using a simplified but realistic transformer topology, analytical evaluation and minimization of the global leakage inductance, in a transformer having split and overlapped primary and secondary windings, are possible without great difficulties.

Two simple but important conditions have been identified for the minimization of the leakage inductance. They are:

- Primary and secondary must be split into blocks and overlapped in such a way that blocks of the same index stay close to each other and show identical absolute values of magnetomotive force.
- Primary and secondary blocks must have identical thickness.

If these conditions are met, then the resulting global leakage inductance is minimized and its value is the one we get without splitting and overlapping, multiplied by a reduction factor close to the squared value of the number of blocks.

References

R. Brault
"Basse fréquence et haute fidélité" Librairie de la Radio, Second edition, pages : 393 to 421.

Norman H. Crowhurst
"Audio Transformer Design" Audio Engineering, February, 1953, Pages: 26 to 27 and 46 to 47.

William M. Flanagan
"Handbook of Transformer Design & Applications" McGraw-Hill, Second edition, pages: 10-6 to 10-7.

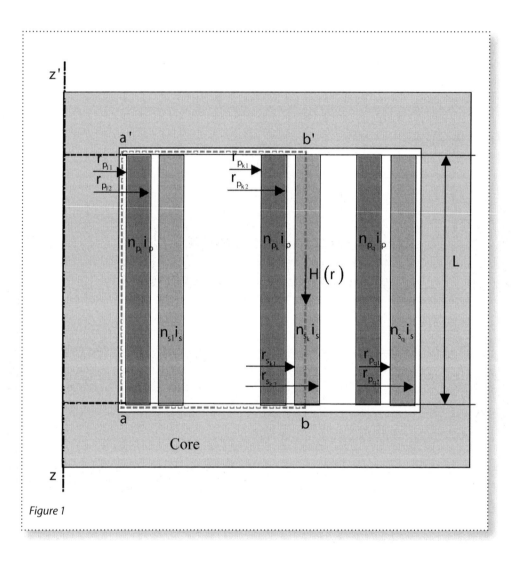

Figure 1

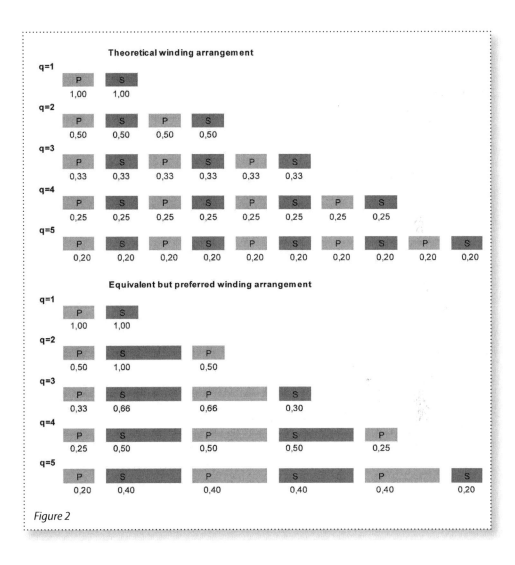

Figure 2

The IC holder notebook

Ed Simon

After spending a frustrating amount of time looking for some DIP IC's I needed for a project, I decided that there had to be a better way to store chips and be able to find them quickly.

This became a simple project that allows the parts to be stored in a smaller space and be easier to find.

I used an ordinary notebook with pages made of standard manila page dividers covered with conductive foam. Mouser sells conductive foam. Part # 809-11250 is a sheet 24" x 36" by a quarter of an inch thick.   This is the cheaper foam; the better stuff shown in the catalog will probably last longer but is a bit more money.

I cut the foam with a razor knife to size and used spray glue to hold it in place. Spray glue is quite messy so be sure to use a fresh piece of newspaper each time you spray a piece.

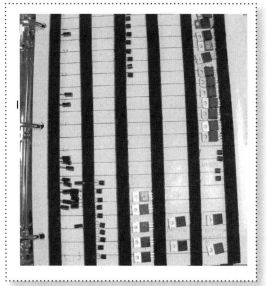

This worked great. I could now store all of my IC's in just 2 notebooks. I then tried using the same method for transistors, but it didn't work out quite so well. Instead I used my computer printer to line the manila page dividers so that I could store transistors sorted by measured ratings. This time spray glue did not seem to hold well enough so I used the brush on contact cement. This allows me to not only store and find transistors faster, but when I get them in I now sort them by Beta and can keep them in sorted order. So matched pairs or even quads are now readily available to me.

Book review - Small Signal Audio Design

Andy Bryner

Small Signal Audio Design - Douglas Self. Focal Press (Oxford, England, UK, www. elsevier.com), 2010. Paperback, 556 pages, ISBN 978-0240521170. $ 62,95

Douglas Self is well known among diehard DIY audio enthusiasts, primarily for his book Audio Power Amplifier Design Handbook. His books have provided much food for thought and fodder for discussion in online discussion forums and elsewhere. He has also written Self on Audio, a collection of articles originally written for Wireless World covering preamplifier and power amplifier design among other subjects. Small Signal Audio Design, his latest offering and the subject of this review, concentrates on design and implementation of small-signal circuitry for home and professional audio applications.

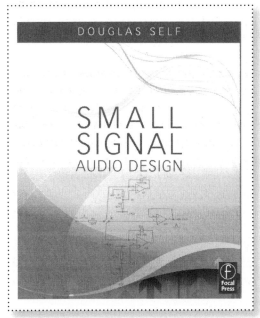

To see how Small Signal Audio Design fits in with Self's previous books, it's helpful to look at their history. Because Self on Audio is a collection of individual articles, it lacks a clear central organization of ideas. Audio Power Amplifier Design Handbook added new material to the original Wireless World power amplifier articles, but more importantly it unified the ideas of those articles, providing a systematic treatment of power amplifier design techniques and details of best implementation practices. Until now, this had not been done with his articles pertaining to small-signal circuitry. That's where Small Signal Audio Design comes in.

Considerable material is covered in the book's 548 pages and 21 chapters. Central themes of

the book include the simultaneous achievement of low noise and low distortion, minimization of cost, the use of negative feedback for distortion reduction, and minimization of crosstalk where applicable. The usage of IC op-amps is heavily emphasized, with discrete transistors only being used for specialized requirements such as ultra-low noise or very high voltage or current swings. These themes are explored in the context of a wide variety of circuit types including general preamplifier circuits, phono preamps, moving-coil head amps, microphone preamps, mixing consoles and filters.

The book begins by briefly describing some fundamentals of negative feedback and noise analysis. The need to keep feedback loop resistors small to minimize noise is emphasized, uncovering a potential trap in real-world designs. An op-amp may feature both low noise and low distortion in its specification, but if low distortion is not maintained when loaded with low impedances such as 600 Ohms, a combination of the lowest possible distortion and noise will not be achievable in practice unless the voltage swing is limited.

One surprise is the inclusion of a chapter on passive components and their effect. Self has acquired an Audio Precision AP SYS-2702 audio analyzer, allowing the measurement of lower levels of distortion than shown in his previous books. Measured distortion of electrolytic capacitors is presented in an interesting way. Since a non-negligible AC voltage across an electrolytic capacitor is needed to cause distortion, choosing the capacitor to be very large should render this distortion unmeasurable in coupling applications. Self examines this possibility and determines a limit on capacitor AC voltage to do just that. Such information is more useful for design than capacitor distortion levels at a fixed AC voltage. Distortion measurements of polyester, polypropylene and polystyrene capacitors are briefly presented as well, showing poor results from polyester and nearly unmeasurable distortion from polypropylene and polystyrene.

The discussion then turns to building blocks for active circuitry in the form of discrete transistors and IC op-amps. Unlike Audio Power Amplifier Design Handbook, Self does not do a deep examination of how to get ultra-low measured distortion using discrete transistor design. With the exception of some unity-gain buffers, the discrete designs presented have measured distortion that's very good, but not quite as good as the very best IC op-amps. Various IC op-amps are subjected to a set of distortion measurements under varying conditions, with the LM4562 showing the best performance.

A comprehensive treatment of preamplifier design is provided. This is introduced by a conceptual chapter describing the system requirements and tradeoffs. The design goals are a combination of low noise and low measured distortion, with an emphasis on achieving the best possible noise performance even with the volume control set for a high attenuation. This latter requirement precludes the use of the common architecture in which a pot is followed by gain block. Instead, the Baxandall active gain control is used. The Baxandall circuit maintains good noise performance as attenuation

increases, and its use of linear pots provides better channel gain matching than alternatives requiring log pots. The chapters covering preamplifiers are very detailed, encompassing moving-magnet phono preamps, moving-coil head amps, volume and balance controls, variable-turnover tone controls, high-isolation selector switches and many other areas.

Mixing console design is treated in a similarly comprehensive manner. To those not familiar with the activities of professional recording and the details of required functionality of mixing consoles, the going gets tougher here, due in part to the associated jargon. The various functions of a mixing console are described in terms of how each one supports each activity in the process. The design emphasis is again on the combination of low noise, low distortion and low cost, with some other surprising design aspects considered as well. Since mixers can process many channels of audio in one package, capacitive crosstalk between traces on the printed circuit board can be a concern. It's shown that when even a small series resistor is added to an op-amp output for stability against capacitive loads, crosstalk at the high-frequency end of the audio band can be degraded by a surprising amount. Self shows that by removing the series resistor and replacing it with an output lead compensation network inside the feedback loop of the op-amp, high-frequency crosstalk performance can be greatly improved while maintaining stability with capacitive loads. Other circuit design topics related to mixing consoles are covered, including high-performance variable-gain microphone preamps with discrete transistor front ends for lowest noise, complex switching mechanisms including electronic switching, and best practices for the design of large-scale summing networks.

Aside from design considerations specific to preamplifiers and mixing consoles, many circuit types useful in a broad variety of applications are discussed. This includes balanced and unbalanced line inputs and outputs having low noise and distortion, interface circuitry for digital-to-analog and analog-to-digital converters, power supplies and a variety of nonlinear circuits. One particularly interesting nonlinear circuit covered is a novel voltage clamp having the lowest possible distortion when not clamping. A chapter on filters is provided, but this very large area is treated only briefly. Distortion in Sallen-Key filters is discussed, but the discussion only treats distortion caused by capacitors. Curiously, the work by Billam [1] on distortion of Sallen-Key filters is not mentioned.

Overall, the book is a satisfying and worthwhile read. Self's writing is easy to follow, giving the reader a solid grasp of the overall concepts and the pros and cons of various design approaches and implementations. Specific design equations are not always provided; it's expected the reader will either already possess the knowledge required for the implementation or consult a general-purpose electronics design text. That approach is consistent with his other books. Abundant measured data are presented, giving the designer good guidance of what to expect in practice. Readers looking for highly detailed coverage of ultra-low distortion discrete transistor circuits will find some good information, but possibly not as much as they'd like. However, given the broad scope of the book compared to Audio Power Amplifier Design Handbook, that's a reasonable tradeoff. Given Self's extensive experience in the field, it's likely that even experienced design engineers will benefit by reading the book.

[1] Billam, Peter J., Harmonic Distortion in a Class of Linear Active Filter Networks, JAES volume 26 number 6, p426, 1978

Book sneak preview - Designing Audio Power Amplifiers

(ed.)

Designing Audio Power Amplifiers - Bob Cordell. McGraw-Hill, New York, NY, September 2010. Paperback, 608 pages, ISBN 978-0071640244. $ 60.00

Sometimes a lot of good things happen together. Morgan Jones brings out updated versions of both of his tube books. Douglas Self brings out his excellent Small-Signal Amplifier Design book (reviewed in this issue by Andy Bryner). And now Bob Cordell has announced his long-awaited Designing Audio Power Amplifiers. Unfortunately, when Linear Audio went to press, Bob's book was not yet out on the streets; it should be by 1 October 2010.

Over many years, Bob has build himself a reputation for being able to explain complex issues clear and understandable and his designs always have an element of innovation and insight. I'm looking forward to get my hands on this book. I'll ask Andy to give it a thorough review in the next Linear Audio. Stay tuned.

Remember your first single?

René Wouda

"Those born in the 1950s are to be envied", a poignant phrase I came across in a collection of essays just the other day[1]. Although it's a sentiment that is often expressed by those that weren't, and with which I – at least I thought so for a long time - completely agree, this sentence kept reverberating in my head. Is it true? Are those sixty something-people to be envied? To me, being born in the 1950s means having had your formative years coincide with the rise and – if you ask me - basically the culmination of modern rock music. Beatle mania being replaced by the more avant-gardism rock music of bands such as The Velvet Underground and The Doors. It's the golden age of television; allowing for the English language programs' invasion in non-English speaking countries. I picture those sixty something's as teenagers in both Europe and the United States crowding in front of the television, eagerly awaiting the performances of new groups, seeing the conventions of pop music being flushed down the drain. Having only listened to the dreadful feel-good songs of innocence that dominated the musical scene on both continents before, growing up in the 1960s must have been mind-blowing. Being born in the 1950s meant you had the right age to enjoy the summer of love and to live the dream of Woodstock. Yet, you were young enough not to be devastated by its sudden ending brought along by Altamont and Sharon Tate. To me the song by the Police sums it all up pretty well.

It must have been a privilege to have had a front row seat in such turbulent times. Bands the like of which we haven't seen since, jamming away during your adolescence. Both my parents have on several occasions recounted purchasing their first single with a sentiment that I can't help but being slightly jealous of. I can't seem to recall my first single, and this is typical. Ask any person of, let's say, 25 to 35 years old about their first single and they all glaze over not understanding. Next, ask any baby-boomer and you'll get a narrative that will go on for at least half an hour. It's all about saving money, spending hours in the local music store, searching for that particular LP and sharing it till late at night with your friends. Today, music is released immediately worldwide. I can't image having to wait for a particular song to be broadcast on the tube or radio or having to wait until I would have

---

1   *Joost Zwagermans - "Perfect Day"*

enough money saved to buy the bloody record. Today you can find anything, immediately, in huge stores, on one of the several hundred radio stations or the internet. I have friends who don't bother to buy music anymore but just create a YouTube playlist. If you do care to have the actual music yourself, it's relatively cheap, especially when you download it from Usenet or a torrent site.

All these changes in the music industry - or maybe not so much the industry but the changing consumers - have made music less exclusive and so much more main stream than it was in the 1960s and 1970s. Wasn't collecting music a time consuming hobby and an accepted intellectual pursuit at that time? You had to actually go to a concert to see a band perform instead of buying a DVD. On the other hand, it's so easy for me to find and enjoy this particular bootleg that otherwise would have taken a true odyssey. I have all my favorite 1960s and seventies music conveniently on one single iPod. Converted to mp3 I'm able to carry 50.000 songs. I guess that's more than an entire music store in the 1960s could offer. Should I forget my player, I still can get my music streaming from any computer or telephone with internet access.

I guess that although I've not experienced such exciting and turbulent musical change during my youth, the development of new ways to appreciate my music anytime, anywhere, is a fair tradeoff for missing the 1960s.

PS I do now remember my first single! My first "single" was actually a tape. A copy of Dire Straits' Brothers in Arms.